T0362725

Urban Beekeeping

Cormac Farrell is an environmental scientist and bee-keeper, best known as the Head Beekeeper for the Australian Parliament. He manages several apiaries throughout Canberra, including training apiaries, organic orchards and tall rooftop apiaries, creating unique food experiences that educate and inspire, pushing the boundaries of what cities can produce.

PRAISE FOR *URBAN BEEKEEPING*:

'This is a thorough, informative and delightful book. Open to any page and you'll find fascinating stories, intriguing bee-facts and beautiful pictures grounded in Cormac's broad experience and deep knowledge.'

— *Stuart Anderson, co-inventor of the Flow Hive*

'Beeautiful! Delightful and essential reading for the beginner and advanced bee lover alike ... Reading *Urban Beekeeping*, you'll get to know Leafcutter, Blue Banded, native stingless, and a swarm of other bees. And you'll get to know the author, Cormac Farrell, whose enthusiasm, intelligence, thoughtfulness and warmth infuse every page. This is not just a book about bees. Woven through the buzz about our honey-making friends is an underlying message about our relationship with nature and our responsibility to shape a future we can share with all living things.'

— *Tanya Ha, environmental advocate and science journalist*

Urban Beekeeping

The City as a Hive

CORMAC FARRELL

First published 2024

Exisle Publishing Pty Ltd
PO Box 864, Chatswood, NSW 2057, Australia
226 High Street, Dunedin, 9016, New Zealand
www.exislepublishing.com

A CiP record for this book is available from the
National Library of Australia.

ISBN 978-1-922539-85-4

Designed by Mark Thacker
Typeset in Minion Pro Regular 11 on 16pt
Printed in China

This book uses paper sourced under ISO 14001
guidelines from well-managed forests and other
controlled sources.

10 9 8 7 6 5 4 3 2 1

This book would not be possible without all of the folks who have encouraged, mentored and even criticized me over the years. However, a special thanks goes to my partner, Lisa, for yelling at me to get the swarm off her favourite yellow dress all those years ago! She is the actual queen bee in my life — love you!

Contents

FOREWORD

It was the spring of 2013, and I had a big problem — my bees had swarmed. Not just anywhere, but onto my partner's favourite yellow dress. She called me to impress upon me (at high volume, and with salty language) how important it was that the bees were off her dress pronto! Normally this is great for a beekeeper — swarms are a chance to establish new colonies — but I had another rule: only one hive allowed at home.

I didn't realize it at the time, but this was a pivotal point that would impact both my beekeeping and my career. The solution was to convince my boss to let me keep the bees on a balcony at our office in the city. I knew that people had been keeping bees in the middle of cities for almost as long as we have had cities. New York City has a world-famous group of beekeepers, as do many Australian cities. I successfully convinced my boss that it was not an entirely crazy idea to have office bees as pets, and shortly afterwards we had a thriving office hive in the centre of Canberra. The bee colonies were a point of interest for the office, and in a strange way put us on the map; we were a small outpost of the larger offices, but suddenly we had company-branded honey jars highly sought after as client gifts.

Along the way I became known as the 'bee guy' in the company, which opened doors and started conversations, helping my team to build networks. It also opened my eyes to how bees are more than just a means to get more fruit and some honey — they are a pathway to changing how we see city environments and urban landscapes.

Then came the really big jump: we had to move offices and the new place had no safe roof access for beekeeping, so we had to cast about for new digs for our homeless office beehives. The Australian National University beekeeping club was in a similar situation, with a new student accommodation building about to be dropped into their apiary site. They had some alternative sites, but they had a promising lead: our national parliament. A recent inquiry had highlighted the importance and vulnerability of bees globally. As luck would have it, the Department of Parliamentary Services was keen to try something outside the box for their sustainability strategy and, after several months of negotiations, we were moving beehives into the Parliament gardens in the dead of night, and I was suddenly the new head beekeeper for the Parliament of Australia. Thus the transformation from backyard beekeeper to the nation's 'bee guy' was complete!

It is a strange thing keeping bees. In many ways they are alien to us, acting more like a distributed 'hive mind' than a collection of individuals. Honeybees send each other detailed coordinates for flowers, using interpretive dance. That would be strange enough, but this dance is done in the darkness of the hive, with the surrounding bees sensing the steps and direction of the dance by the vibrations it sends through the honeycomb. Native stingless bees are just as ingenious, leaving a scent trail as they fly to allow their sisters in the hive to home in on flowers they have found.

Bees have fascinated people for millennia. Whether it be due to a casual use of honey on toast, their role in agriculture or the religious and mystical inspiration bees create, people have always loved bees. When many people think of bees, they immediately think of the ubiquitous honeybee, *Apis mellifera*, which underpins much of our agriculture around the globe. However, the actual role of the over 20,000 species of bees worldwide is now becoming increasingly apparent, not to mention all the other pollinators we rely on.

The one thing that all bee species are dependent on is abundant flowers that are free of pesticides, and this drives the most important change in your mindset that you have to make to be a successful beekeeper. Bees are intrinsically connected to the landscape that surrounds them, and as a beekeeper you will see the world through a different set of eyes. What previously looked like empty space or waste ground now looks like an opportunity for a flower garden. For me, this meant that the trees (which I already loved) became even more valuable if they flowered — these are the powerhouses of urban food production and habitat for all sorts of species. It also became critical for me to influence home gardeners to be thoughtful and careful in what they spray, lest this be carried back to hives in the surrounding suburbs.

One of the things I love most about my original career as a forester was how small and insignificant you feel when standing in an ancient forest. You come to realize that your time and place on this planet, for all the seeming importance that we get wrapped up in day to day, is really nothing. Bees tend to have a similar effect on me — I find it endlessly amusing (and humbling) that for all of our magnificent technology, literature and art, we are almost completely reliant on bugs in a box to feed ourselves.

One of the central conceits that beekeeping explodes is the idea that humankind is somehow superior to the other creatures on this planet, that other creatures are expendable in our quest for development, profit, life in general. The reality often dawns on new beekeepers fairly early: it is in fact the other way around. We are the expendable ones, and it is plants and their pollinators, soil microbes, and detritivores that are the really critical part of life on Earth. This flip in perception often changes people's view of the world in a fundamental way.

When I was training at university to be a forester, we had an apiarist (as bee farmers are formally known) come in to talk about the value that forests held for bees and beekeepers. It was a revelation, particularly how complex proper nutrition for bees was, with the plants on the forest floor (that foresters too often ignored) in fact being critical for sustaining hives. I carried this lesson with me until many years later, when it became a critical part of designing sustainable urban forests for pollinators.

I want this book to be part instruction manual for urban beekeepers, and part manual for city planners, governments and citizens in creating a more sustainable way of life. When I was growing up there was a very clear sense that cities were for people, with green spaces to be carefully controlled, to look pretty but be basically useless. Wildlife and wild things were to be left in the bush. It always seemed a bit dull to me — why can't we have some wild things in town? And why can't we grow at least some of our food in our cities?

Of course, this means that we have to change the way we live, and the way we design our urban spaces. We have to be open to less concrete and more green, less control and a bit more give towards nature. In return we get cleaner air, more green space to help keep us sane, and delicious, unique food to be proud of. Seems like a reasonable trade to me.

1

Bees in the city

Subverting the Concrete Jungle

When I was growing up, there was a stark divide between the city and the country. I always knew which of the two I preferred. We lived close to the edge of the city, and I spent a lot of my early teenage years running traplines for rabbits and foxes, fishing in the local streams and generally being outdoors every chance I got. Later when I started working as a forest ecologist, cities were pretty much given up as 'dead' land, with the focus of conservation and land management 'out there' in the wider countryside.

More recently, though, there has been the realization that we don't have to have a hard delineation — we can bring some of the natural landscape into the city. Not only does this bring aesthetic beauty and wildlife into town, but increasingly we are realizing the practical benefits of this. Urban areas with higher tree canopy cover consistently outperform other areas for property prices, and improve health outcomes. The increasing prevalence of heat-waves has highlighted the importance of shade and cooling vapour from the trees as they breathe (yes, trees actually do breathe in a cycle).

However, the thing that I have been most passionate about has been the capacity for the urban forest to provide food. Not just for us, but for wildlife that is increasingly relying on cities as refuges and transit paths between remnant habitats. Of course, this can never fully

The Parliament bees happily foraging on the River Peppermint trees fringing the entrance roads.

Amazing where you can cram a bit of greenery into urban areas — like under this powerline.

substitute for natural habitat, but it definitely helps, and allows us to enjoy some unique food and beautiful wildlife along the way.

When I first started urban beekeeping, one of the funniest episodes was convincing the lawyers from the company I was working for that it was even possible to keep bees in the city. 'What will they eat? Surely they will starve?' they asked on several occasions. It is a reasonable question — for so long the central business districts were a concrete jungle, devoid of green life. Happily, there have always been gardeners sneaking greenery into the city, and this has now accelerated as attitudes change. In reality, our towns and cities are some of the best places to be a beekeeper and produce honey to rival the best in the world. This was illustrated in a funny anecdote one of our company lawyers later relayed to me. At their annual meeting with their insurers in London, they had prepared a detailed brief on the company beehives. The insurers were completely unfazed, apparently commenting 'Oh, you mean like the hives the Queen keeps at Buckingham Palace?' Turns out that the Queen had her own urban hives producing honey in the heart of London, and the idea of honey from the city of London was a long-established tradition, seen to be a completely normal thing to do.

Keeping bees on public buildings and landmarks is increasingly the normal thing to do, and has been happening in European cities for some time. Paris has hosted beehives on landmark buildings for decades, with Louis Vuitton headquarters, the Paris Opera and Notre Dame Cathedral all supporting thriving colonies of honeybees. The Notre Dame bees even miraculously survived the 2019 fire that destroyed much of the building, and have become a symbol of hope during the rebuild.

Across the Atlantic, the White House started keeping bees under President Obama, and these bees were a focal point when he established some of the first pollinator protection policies. While his successor was certainly no environmentalist, his vice president, Pence, is a keen beekeeper. Under President Biden a new Pollinator Protection Initiative has seen beehives placed on government buildings at eleven locations across the United States.[1] The United Nations headquarters in New York has a beautiful Slovenian AZ hive, which is a pair to the one installed at the Slovenian embassy in Canberra.

Here in Australia, the Parliament of Western Australia became the first to keep bees, with the Queensland Parliament keeping a pair of native stingless bees. Both our former and current Governors-General are patrons of World Bee Day, with Government House in Canberra supporting several hives. Their boss, Charles III and his wife Queen Camilla, are also well-known beekeepers, and are patrons for the Elephants & Bees charity that supports beekeeping for poverty alleviation in Africa. Surely a royal decree to care for our bees across the nation cannot be too far away!

This dense bank of Lavender provides beauty, wonderful scent and constant nectar for one of my nearby rooftop apiaries.

BECOMING AN URBAN BEEKEEPER

For me, the transition from a backyard enthusiast to something more serious was a gradual process, a case of seeing opportunities, sometimes biting off more than I could chew and then chewing like crazy! However, the common element was always fitting the approach to beekeeping to the space that I had available. Sometimes this meant establishing honeybee apiaries, while in other situations native bees were the best choice.

The great news is that the estimated 2000 species of bees that we have in Australia are some of the most adaptable, useful and beautiful creatures that live alongside us, and almost anyone has some space for them. The space that you have will largely define what you can do, and may even go so far as to choose the species of bee that you keep.

Most of us start as a beginner beekeeper with one or two hives in our backyard, and that is certainly how I started with honeybees. Moving into an urban or inner-city space is different: you potentially have the freedom to have more hives than you ever could in a suburban backyard, but at the same time you are often presented with a blank canvas — either an empty field, a concrete rooftop or a patch of waste ground that nobody else wants. Equal parts daunting and exciting for most of us!

Moving a hive of live bees to the roof in a lift. Totally normal, right?

Hanging out in one of my tall building rooftop apiaries.

Getting close up with one of my bees

Generally, urban spaces where bees are kept tend to fall into two broad categories: open areas at ground level that can often work like a miniature farm, and apiaries crammed into the spaces of a large structure, often a roof, balcony or even the spaces between buildings. Some urban beekeepers even use a network of suburban backyards to spread small groups of hives across the wider landscape. Each situation offers its own opportunities and challenges, but in almost all cases it is possible for bees to survive and thrive!

Most importantly, all such beekeeping situations provide opportunities to change the way our communities see the urban landscape, often in a way that opens up opportunities for food, friendships and fun.

2.

Meet the bees

Native bees and honeybees

SOLITARY NATIVE BEES

Often, when people hear the word 'bee' they immediately think of 'honeybee', which has become the default bee worldwide, mostly because almost everyone has tried (and loves) honey. Honeybees are important, with honey and other hive products generating around $100 million per year in Australia. The contribution of honeybees to agriculture through pollination services is estimated to be even greater, valued at around $14.2 billion in Australia in 2017.[1]

However, honeybees are just one of an estimated 20,000 bee species that we share the planet with (and they're not even the only bee species that produces honey). The majority of bee species live relatively solitary lives and produce no honey that we could ever enjoy on our toast. However, any lack of interest in them is misplaced, as they are in fact responsible for a large portion of the food we enjoy. While honeybees are critical in agriculture for pollination, it is becoming increasingly apparent that, for many crops, a significant portion of the workload is carried on the backs (and legs and bellies) of native pollinators. Just for bees alone, Australia has an estimated 2000 species that pollinate, and

A beautiful native Spring bee (*Tricholletes*) on a native Pea Flower. (Credit: Dr Kit Prendergast)

Peekaboo! An adorable native bee peeking out of its nest. (Credit: Dr Kit Prendergast)

Bee hotel on the main street of Hall, a bee-friendly village!

that is before we get to butterflies, moths, flies, beetles, bats and birds.

Any discussion of bees in Australia needs to start with our natives, as they are too often overlooked but are an important part of a functional landscape. Going through the full list of our native bee species would fill an entire book on its own, and their stunning beauty has seen them become increasingly popular on social media. It is worth discovering the native bees in your area and making some space in your garden for their specific habitats. Some are even incredibly efficient pollinators, in many ways superior to honeybees. Spread some wood blocks, mud bricks and hollow reeds throughout your garden and you will have the potential to build a haven for biodiversity — and a diverse mini-army of pollinators for your flowers and vegetables.

There is another reason that we need to start with native bees, especially if we are trying to encourage sustainability: almost anyone can create a habitat for them. Not everyone wants to keep honeybees, and not everyone should. Native bees still engage people in their local landscapes and environments, and in some cases more so. Most native bee species have a small home range — around 200 to 800 metres — so the bees you find in your backyard really are 'your' bees. Recent research has found that high proportions of native bee-friendly flora beget high native bee biodiversity.[2] It is the classic 'think global, act local' thing that we can all do.

This also makes native bees much more susceptible to the 'pollinator deserts' that beset many urban areas; both a physical break in suitable flowers, as well as a seasonal gap in forage can severely limit their populations. This is one situation where actions at a local level can have a really big impact on conservation and landscape biodiversity.

The bees themselves are pretty cute, too, so with a bit of planning you can turn almost any urban space into a mini-menagerie of adorable pollinators.

Garden headbangers: Blue Banded and Teddy Bear bees (*Amegilla* spp.)

If you like tomatoes, you have to love your *Amegilla* species, the Blue Banded bees and their adorable cousins the Teddy Bear bees, as these beat honeybees hands-down for pollinating all of your *Solanaceae* vegetables (such as tomatoes and eggplant). These plants have developed a clever way to prevent their pollen from being wasted on wind or a generic pollinator, and instead rely on something called buzz pollination. The flowers keep a tight hold on their pollen until they feel a shake at just the right frequency, which the bee creates by grabbing the flower and running its wings in neutral, called 'sonication'. The resulting buzz is quite loud, and on a calm day it sounds like the bees are using a mini taser on the flowers. By banging their heads against the flower at the right pitch, like a secret password, a shower of pollen is released onto the bees.

In the Australian environment there are native *Solanaceae* (e.g. 'bush tomatoes') as well as other plants that use buzz pollination like Strap-Lilies (*Dianella* sp.) with tomato-like flowers.

As a predominantly ground-nesting bee, *Amegilla* species prefer some soil to burrow into, ideally where other bees of the same species are hanging out. In urban areas

Blue Banded bee (*Amegilla*) — one of our most beautiful and valuable pollinators

Mud brick bee hotels are easy and fun to make — and very popular with Blue Banded bees!

Blue Banded bee doing what it does best — and being beautiful at the same time! (Credit: Dr Kit Prendergast)

they sometimes work their way into the mortar between bricks in walls, but for artificial nesting habitat mud bricks are the best way to create your own native bee high-rise. As a communal-nesting species, they want a stack of mud bricks, ideally ten to twelve high. Their main nesting habitat in nature is vertical creek banks, and the vertical stack of mud bricks needs to emulate this.

While they mostly have their own space in their in-ground burrows, they prefer to live in loose collectives. And while not quite as organized as the hives of communal bees, Blue Banded and Teddy Bear bees form collective groups from a few bees to several dozen. The males tend to be the easiest to spot, as they hang out together on bushes above wherever the female bees have made their burrows.

Saving your stems for *Allodapini*: Reed bees

Two of my favourite things to grow are Raspberries for their delicious fruit, and Sunflowers for their beauty. Both also produce nectar and pollen in abundance. Once the plants are finished, you are often left with a bunch of hollow stems — don't throw these out! They are perfect for creating habitat for one of our smallest groups of natives, the *Allodapini*, known commonly as Reed bees. These tiny native bees nest in the dried-out hollow stems of plants, and this is how you attract them to your garden. They don't need fancy accommodation; simply saving your old Raspberry canes, sunflower stems and other hollow

Reed bees (*Allodapini*) love the hollow stems of plants for their nests, and visit a wide range of flowers in your garden.

The stunning Metallic Green Carpenter bee, one of our iconic and increasingly threatened native bee species. (Credit: Dr Kit Prendergast)

plants, tying them into bundles and hanging them throughout the garden is enough.

They are small bees, generally 5 to 8 millimetres in length, and they have a flattened, wedge-shaped tail section that they use to block the entrance to their nest when they are home.

Xylocopa: Carpenter bees

If you have excess wood that needs chewing, then *Xylocopa* — Carpenter bees — are the bees for you! They are a little bit picky in their choice of wood but are among the largest and most colourful bees you can attract to your garden. I remember seeing a Metallic Green Carpenter bee (*Xylocopa (Lestis) aerata*) for the first time; it didn't look real, as if someone had made a robot bee out of emerald-green metal. They are also a large bee and forage on a range of plants, especially ones that require buzz pollination.

These bees really prefer softer wood to build their nests, so you need to save some specific types of wood for them to build their nests in. The classic is the long flower stems of the Grass Tree (*Xanthorrhoea* sp.), which have a soft centre perfect for tunnelling into.[3] The main downside is that these plants take several years to mature and will only occasionally flower. When they do, though, definitely save the stems and ideally leave them in place; they are a fantastic resource for any Carpenter bees in your garden.

Artificial stems and blocks of balsa wood are used as part of bee conservation projects, especially when placed among plantings of *Xanthorrhoea* or Banksia, which form the Carpenter bees' favoured habitat after low-intensity fire in natural bushland.[4] Fire is not often encountered in an urban backyard, so ensuring your garden has abundant native flowers and some artificial nest stems is your best bet.

In northern areas of Australia, the Great Carpenter bee (*Xylocopa (Koptortosoma) aruana*) starts to make an appearance and as our largest bee, with striking black and yellow markings, it is pretty hard to miss! As with their metallic green cousins, they favour soft wood that they can carve into with their strong mandibles, so blocks of aged Banksia trunk or blocks of balsa are your best bet. Mostly they tend to find their own pieces of soft wood to burrow into, perhaps in a fence or a shed. If you see them nesting in these places, retain that wood, as they are semi-social and will form aggregations that live near each other.

THE *MEGACHILIDAE*

Megachile (Eutricharaea) sp.: Leafcutter bees

These are among the cutest of the native bees. They are generally around half the size of a honeybee and one of the most common inhabitants of home garden bee hotels. Their name comes from the female's habit of cutting sections of leaf from nearby plants to create a nest tube, partitions and door, in which the nectar and pollen, and developing larvae, are encased. Leafcutter bees seem to be semi-solitary in my experience, forming loose groupings around a nesting site. In nature, they use the holes in trees created by wood-boring insects, but in garden bee hotels they will happily use holes drilled in wood blocks or hollow bamboo stems. There are lots of different species that come in different sizes, so it is worth having a range of nest hole sizes for them to choose from. Some species will also nest in the ground.

In addition to being popular and adaptable for bee hotels, Leafcutters are really fantastic pollinators. Unlike honeybees, which carry pollen on their legs, these species carry pollen on their belly, creating a colourful display as they forage. The pollen often drops from their belly as they move around the flowers, creating excellent pollen transfer between flowers. Leafcutters visit a wide range of native and exotic flowers, but when cutting leaves they seem to particularly love Roses, probably due to their soft and easily folded leaves.

When preparing a bee hotel for Leafcutters, two small details are important. First, the old holes left by wood borers have smooth sides, and these bees just let their wings slide

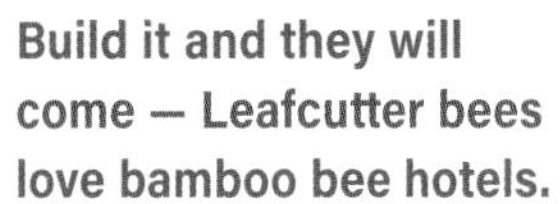

Build it and they will come — Leafcutter bees love bamboo bee hotels.

Leafcutter bee (*Megachile*) enjoying the Golden Rain Tree blossoms in my garden.

along the walls without fear. If the holes drilled in the wood are not smooth (achieved by using a high-quality tungsten drill bit) then the rough edges can tear the bees' wings apart. Second, the holes need to be deep, at least 20 cm, as shallower holes produce males that hatch first, with the deeper sections hosting female larvae.

The other *Megachilids*: Resin bees

Where Leafcutter bees use the folded-up leaves of plants, some *Megachilids* — Resin bees — collect resin from plants, using this to create partitions between their brood cells and cap their nest. Resin Bees are part of the same genus as Leafcutter bees, and they use bee hotels and your garden in a similar way. Due to the resin, bee hotel holes that they fill with their brood look really distinctive, often with a red resin cap that stains the surrounding wood. The bees themselves are also really beautiful, and one of my favourite bees is the Fire-tailed

Resin bee (*Megachile mystaceana*). As the name suggests, this species has a spectacular orange/red abdomen, making them really stand out in the garden.

Ground-nesting bees

There is a wide variety of ground-nesting bees throughout the world, but my personal favourite is the *Lasioglossum* species that frequents my patch of native Bulbine Lilies (*Bulbine bulbosa*). The males of these cute little bees have a habit of sleeping inside the flowers, providing a very social-media-friendly photo opportunity. Their real value is pollinating my native plants, where they are constantly working most of my tubular-style flowers. The Bulbine Lilies are a delicious indigenous vegetable, so I am always happy to see them at work pollinating the seeds for my next crop.

The resources for these bees are as simple as it gets — lots of flowers, and leave some patches of bare soil for them to nest in. Seriously, that's it. The females often sleep in the burrows that they make, leaving the males to cluster outside, so if you are observant you might see groups of bees sleeping on twigs or grass stems. They generally want to be close to where the females are, so this is a pretty good indication of where they are building their

Resin bee collecting for its nest. (Credit: Dr Kit Prendergast)

An adorable *Lasioglossum* curling up to sleep inside a flower.

burrows, and where you should leave patches of bare ground. The entrance holes depend on the species, but can sometimes be quite easy to see, clustered together with small mounds of fresh soil at the entrance.

Cuckoo bees

While most bees are gentle vegans, collecting pollen and nectar to raise their brood, there is a group of bees that has embraced the thug life, turning to crime over hard work. In the same way as Cuckoo birds steal the nests of other birds, Cuckoo bees sneak up on their hardworking colleagues, laying their egg next to their victim's baby.

Cuckoo bees might be sneaky thieves but they have one saving grace: they are among the most beautiful bees you will ever see. The Neon Cuckoo bee (*Thyreus nitidulus*) targets Blue Banded bees and is probably the most striking bee in Australia, with spectacular metallic blue and black patterns all over its body. The related Domino Cuckoo bee (*Thyreus lugabris*) targets Teddy Bear bees and is almost as beautiful, with brilliant white spots over a black body.

Cuckoo bees tend to forage on similar flowers to Blue Banded bees and Teddy Bear bees,

Other natives include this Banksia-loving Masked bee (*Hylaeus alcyoneus*). (Credit: Dr Kit Prendergast)

Stunning colours of the Neon Cuckoo bee (*Thyreus nitidulus*). (Credit: Dr Kit Prendergast)

Beautiful but sneaky.

Stingless bees proved to be a hit in the Australian Parliament.

so if you have *Amegilla* species in your garden in sufficient numbers these bees are probably not far behind. In addition to the *Thyreus* that target *Amegilla*, there are also *Coelioxys* that target *Megachilids*, *Sphecodes* (with the creepy common name 'blood bees') targeting *Halictids*, and *Inquilina* targeting the *Allodapini*. Basically, thievery is everywhere, even if you are a bee.

Ironically, if you have these sneak-thieves showing up in your garden then it is a sign that you are on the right track, as they need reasonable congregations of their hosts in order to successfully breed.

COMMUNAL NATIVE STINGLESS BEES

One of the unique things about the Australian Parliament apiary is that we keep not just the usual honeybees, but also our native stingless bees. About a year after we had established the honeybee hives, we received an inquiry from one of the Members of Parliament. A few of his constituents had recently won a design competition with an innovative new type of hive, built specifically to take advantage of our native stingless bees. It was the start of the next phase of Parliament beekeeping, where we expanded into our native honey producers, native bees from the *Tetragonula* genus.

There are eleven eusocial (colony-forming) species of bees in Australia that we know of so far, forming hives of several thousand workers, and containing one mated queen laying eggs, and several virgin queens waiting in reserve. Of these, three species are commonly kept in managed hives: *Tetragonula carbonaria*, *Tetragonula hocksonii* and *Austroplebeia australis*. Surveys by the Australian Native Bee Association have indicated around 1500

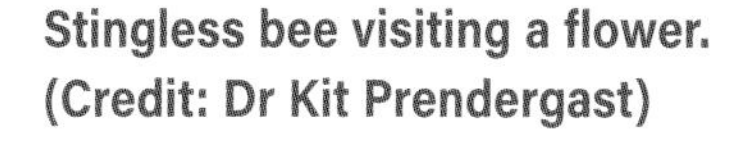

Stingless bee visiting a flower. (Credit: Dr Kit Prendergast)

Beautiful native stingless bees, eagerly awaiting the chance to dismember the next intruder.

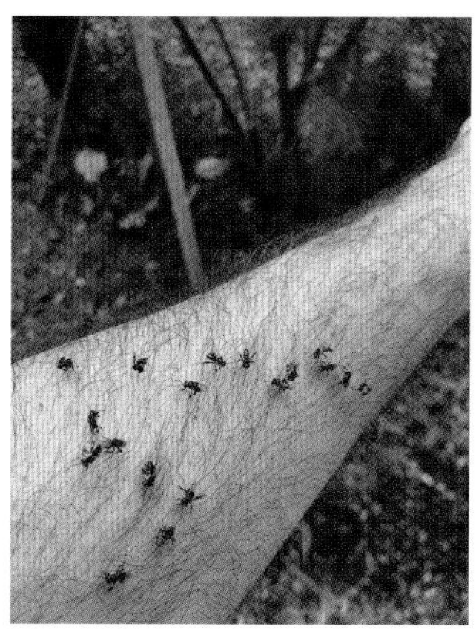

Getting a free hair removal treatment, courtesy of the Parliament stingless bees.

native stingless beekeepers with approximately 12,000 hives. Compared to the nearly 30,000 honeybee keepers in Australia, this is a smaller part of the beekeeping community, but the beauty and ease of caring for native stingless bees means that they continue to grow in popularity.[5] There are even commercial operators now, which has increased the supply of colonies and hive types, making it easier to get started.

Normally this tribe of bees is restricted to the warmer parts of Australia, and simply can't survive the cold winters of our nation's capital. However, these new boxes made by Hive Haven were heavily insulated and designed to be easily transported to migrate the colonies long distances. It gave us an idea: what if we could keep them at Parliament in spring and summer and give them a winter holiday somewhere warmer? As it happened, the New South Wales Governor at the time was a keen beekeeper, and New South Wales Government House was within the natural range of the species.

After a few months of emails and phone calls, and some additional paperwork, we had added a cute new species of bee to our growing apiary!

The stingless bees call Parliament home for the warmer months of the year, generally late spring to early autumn, but as soon as the Canberra cold starts to bite, it is time for them to head off for their winter holiday. The ease of transport for these colonies is a huge plus — honeybees hate being moved, and will get really pissed off, whereas the native stingless colonies seem to be fairly chilled out about the whole thing. Once they are all in for the night, they are sealed with a simple piece of breathable cloth, and then loaded into the car for the trip to Sydney. The colonies are also so much lighter and smaller; you can just pick them up.

The big advantage of native stingless bees is that they are, of course, stingless. You can get really close to them, and this allows a level of interaction with the public that is not

The weird but strangely beautiful interior of a native stingless hive.

Native bee honey is formed in small bags of resin, giving it the name 'Sugarbag'.

really safe with honeybees. That is not to say that they are completely defenceless — they don't sting, but if the colony feels threatened when you are inspecting them, they will try to swarm onto you and pull out any hairs they can get hold of. Surprisingly strong, they are able to lever out arm hair, leg hair, eyebrows and even nose hair (really quite awful). In one memorable incident, they decided to really go after the Minister for Resources, Madeleine King, as she was helping me with an inspection. She had to rush back to Parliament question time, so her staff and I had to get all of the cranky defenders off her pronto! Everything in Australia wants to kill you. Some animals just can't!

Generally, though, these are lovely, gentle bees who don't bother anyone. This makes them perfect for public displays and corporate installations. The colonies are light and easy to transport, and people can get right up close to watch the foragers come and go. The company that supplies the Parliament bees also provides colonies to a range of businesses, creating fascinating living displays. This really proves that it is not all about the honey; despite their modest levels of production, these native bees continue to spark fascination and engagement wherever they are placed.

In nature, most of their natural enemies are other insects, so their home defence is simple but brutally effective. A group of them will swarm onto an intruder, each grab a leg and then pull in different directions, literally disarming their opponent. Even much larger insects like honeybees and assassin bugs are no match for them.

Native bee honey — low volumes but a unique, tangy flavour!

Keeping native stingless bees remains a work in progress in Australia, and colonies can still be a bit hard to come by in many areas. New colonies are most commonly created by splitting an established stronger colony, where the brood is cut or gently pulled in half, taking some honey and pollen stores with each. The kinder but less reliable method is budding, where a new hive is attached via a tube that encourages the 'mother' hive to create a new 'daughter' hive as an outpost of the main hive.

A special part of native stingless bees is their honey — it is unlike honeybee honey in many ways. The first thing that strikes you is how it is stored by the bees. Unlike the standard honeycomb, the pots of honey often referred to as 'Sugarbag' are stored in bags of resin connected by a crazy network of tubes. This mixture of resin and wax infuses the honey with a tangy taste, but this is only half the story. The tang is actually the antibacterial compounds hitting your tongue, with Sugarbag having equivalent antimicrobial activity to the famous Mānuka honey.

Chemically it is also totally different to honeybee honey, with the stingless bees converting sucrose into a special sugar known as trehalulose. In addition to having a refreshing citrus tang, this honey will not make your blood sugar spike, making it much safer for diabetics to consume. The tang comes from the way the bees store the honey. Unlike the wax honeycomb inside a honeybee hive, their beautiful honey pots are mostly made of plant resins, which they transform into a sticky, rigid glue called propolis. The plant resins sufficiently mix with the honey to give it the unique tangy flavour as well as microbial properties.

All of this comes at a cost, though. Stingless bees' small colonies and tiny size mean that production is a fraction of what a honeybee colony can produce: somewhere between 500 grams and 1 kilogram for most northern areas, and basically no surplus for colonies south of Sydney.

Honeybees attending to their queen.

HONEYBEES

When it comes to producing large quantities of liquid gold, there really is only one species of bee to turn to — *Apis mellifera*, better known as the honeybee. This species has been revered, admired, exploited and even worshipped for millennia. Ancient Egyptians were the first civilization recorded to use migratory beekeeping, moving hives on barges down the Nile to pollinate crops, and revering the honeybee as the tears of the sun god Ra.

These days, honeybees form a critical part in our agricultural systems, but in an urban setting they are mostly maintained to provide local varieties of honey and to pollinate home gardens. Keeping honeybees used to be a bit of a niche hobby, confined to keen gardeners and old, bearded men when I first got into it. However, something changed: a perfect storm of environmental concern, enthusiasm for local foods and improved technology saw a resurgence in beekeeping as a hobby. Suddenly, local bee clubs went from having a few dozen members to literally hundreds. The club that I was president of went from fewer than a hundred members to over 500 in just a few years. It was a bit overwhelming, with hundreds of new beekeepers suddenly needing to be trained. However, it was also quite lovely, with increasing numbers of young people and especially women finding a fascination with bees. The fear that the new beekeepers would be in it only for the honey proved to be unfounded — almost everyone just loved their bees.

Parliament honey jar under the Parliament flag.

3

Getting started

Moving out beyond the backyard

If you are planning to become an urban beekeeper for a community garden or an urban space (like a rooftop), I am going to assume that you have already been successfully keeping bees at home. Yes, it is possible to do urban beekeeping as a complete beginner, but due to the safety risks to yourself and the public I would recommend against it. Given the complexity of managing honeybees in a public space, it is better when you have at least a few years' experience. This book is therefore mostly going to focus on the additional skills an urban beekeeper needs above and beyond a standard backyard enthusiast.

Apiaries with a view: Tall building beekeeping in the heart of Canberra

In the second half of 2018, beehives started appearing on rooftops across Canberra, steadily growing to three city rooftop apiaries, two of which I manage. It was a sustainability initiative of the ISPT Group, which manages a large portfolio of commercial buildings, and has been a fruitful and enjoyable partnership. One of the buildings where I manage the bees overlooks

One of my rooftop apiaries in the centre of the city.

Honey jars from the roof of the building.

Parliament and Black Mountain, and the other is in the central business district of Canberra, where my bees pollinate the street trees and gardens in the city centre.

As a beekeeper, ISPT have been great to work with — I have always had great communication with their staff, they have shown an interest in beekeeping and have come up with a range of innovative events covering both honeybees and native bees. The apiaries each produce substantial amounts of delicious honey for sale to the buildings' workers, and often sells out in less than a week. The workers have enthusiastically embraced World Bee Day, creating experiences like rooftop hive tours and running native bee hotel-building workshops. These tours have been the real highlight, allowing people to experience the stunning views from the rooftop apiaries and learn about the management of the bees. This includes beelining — standing behind the hives and watching where the streams of workers are headed, allowing us to guess which flowers they are foraging on.

The level of communication around management has also been a highlight — a well-developed risk management plan has allowed us to predict when workers are likely to be too close to the bees, requiring the hives to be moved or temporarily closed up. Potential threats such as spraying of the external building or gardens have also been notified well in advance, and we have managed to find ways to avoid any impacts on the honeybees.

The company also created a portable display stand for the honey sales, which can be disassembled and stored between events. This has proved to be a good innovation, allowing quick 'pop-up' sales of honey to occur whenever sufficient honey stocks coincide with pollinator-focused days like World Bee Day or Pollinator Week. The combination of photographic panels and information on bees has been a great way to communicate the importance and diversity of bees to a wider audience.

This beautiful display was created to allow for pop-up pollinator events

LEGAL REQUIREMENTS

The big difference between keeping honeybees and native bees is in the legal space. There are basically no requirements for keeping native bees — it really wouldn't make sense, as they are all around anyway. Honeybees are different: everyone is focused on the stings but disease control is equally important. For most urban beekeepers it is a bit of fun, and perhaps a small side business, but for our agricultural industries' biosecurity it is a matter of survival.

One of my roles with the ACT government is to help them with urban bee problems like abandoned and problematic hives, so I get to see the most common issues that they are dealing with. Most jurisdictions have established a code of practice, and while these are often not completely relevant for community gardens or rooftop apiaries, there are some common elements you should try to follow when setting up an urban apiary:

- **You must always have the consent of the landowner where you wish to place your hives.**
- **You must provide a dedicated water source for the bees.**
- **You should keep a docile, non-aggressive strain of bees (see p. 29).**
- **You must control swarming (see Chapter 7: Hive Management).**
- **You should use escape boards to remove honey where possible.**
- **Be considerate of others when working with your bees.**

Below is a state-by-state outline of some of the basic beekeeping guidelines and the codes that govern them.

Location	Number of hives	Minimum distance from property boundary	Other requirements
NSW *Beekeeping Code of Practice for NSW*	Small block: 2 hives Average block (up to 1000 m^2):4 hives Large block: 8 hives	Bee flight path to be above 2 m when crossing property boundaries	Hives should not be located in the vicinity of schools, childcare centres, hospitals
ACT *Code of Practice for Beekeeping in Residential Areas*	Small block: 2 hives Average block (up to 1000 m^2): 4 hives Large block: 8 hives	Minimum 4 m from access pathways and front doors	Beekeepers are responsible for catching swarms originating from their apiary

Location	Number of hives	Minimum distance from property boundary	Other requirements
Victoria *Apiary Code of Practice*	Ranges from a single hive (up to 500 m^2) through to up to 60+ hives (1 ha and above) Higher densities of hives allowed for short periods to support urban agriculture	Minimum 3 m, unless there is a wall, fence or tall vegetation to direct bees upwards, when it can be 2 m	Beekeepers are responsible for catching swarms originating from their apiary
Western Australia *Best Practice Guidelines for Urban Beekeeping*	Ranges from 2 hives (up to 450 m^2) through to up to 40 hives (4000 m^2 and above)	Minimum 3 m. Bee flight path should be above 3 m when crossing adjoining properties and public paths	Includes recommendations for health, safety, swarm control and transport of hives
Tasmania *Code of Practice for Urban Beekeeping in Tasmania*	No hives for blocks less than 400 m^2 400–1000 m^2: 2 hives 1000–2000 m^2: 5 hives 2000–4000 m^2: 10 hives.	Minimum 3 m, unless next to a barrier or fence of at least 2 m high Hives oriented to have bees fly across the property	Developed by Tasmanian Beekeepers Association
South Australia *A Guide to Beekeeping in South Australia*	Less than 450m^2: 2 hives 451–1000 m^2: 3 hives 1001–2000 m^2: 4 hives	Bees to be kept 20 m from boundaries or behind a barrier at least 2.1 m high	Special quarantine requirements for Kangaroo Island Swarm traps must only contain undrawn foundation Must keep biosecurity records and notify when selling hives
Queensland	No state laws on number of hives; some local council restrictions	No state laws on location of hives or how close to boundaries	Guidelines on reducing nuisance provided, but no specific restrictions
Northern Territory	No restrictions	No restrictions	No specific code of practice

As you can see, there is a large variance in the codes of practice, and my personal experience helping draft the code of practice for Canberra urban beekeepers was an eye-opener. People had a range of dark conspiracy theories about what the government was trying to do with the code, and there was a lot of push-back across the board. Beekeepers tend to have 'interesting' personalities, and I can understand that some jurisdictions would wait until there is an actual problem before poking the beehive and developing a code of practice! Even now, most of the codes are voluntary or enforced though negotiation and mediation, rather than being a set of hard rules. The limits on numbers of hives really only apply to residential areas, and don't really apply to city rooftops and community gardens in outer areas of cities; but they still provide a useful guide. One of the common elements is the requirement to maintain 'docile' or 'quiet' strains of bee — but what does this actually mean? In almost every case, it means changing the queen.

QUEEN BEE SELECTION

The queen is a critical part of the colony, laying the eggs that replenish the workforce, but her name is a bit of a misnomer as it implies that she is in charge. In reality, no single bee is ever in charge of the colony and it is the workers themselves who make all the major decisions. In the case of the queen, the workers decide if she needs to be replaced and a new queen raised. This is the truly mind-bending part of working with superorganisms like bee colonies — no single bee is in charge and yet everyone does their job. In addition to laying the fertilized eggs that grow into worker bees (up to 2000 a day in peak times) the queen bee is an important part of the temperament of the colony, and this is how we maintain docile honeybee colonies to meet the code of practice requirements. Her genetics set the mood of the hive and will determine things like how defensive the colony is; even some disease-resistant traits are determined through her breeding.

Three main breeds of queen are sold in Australia, and if you are buying from a registered breeder, they will be able to provide a lot of information on the traits they are selecting for. For urban beekeepers, the most important trait is gentleness — bees that are not defensive when people go near the hives, and do not become aggressive when inspected.

Italian queens

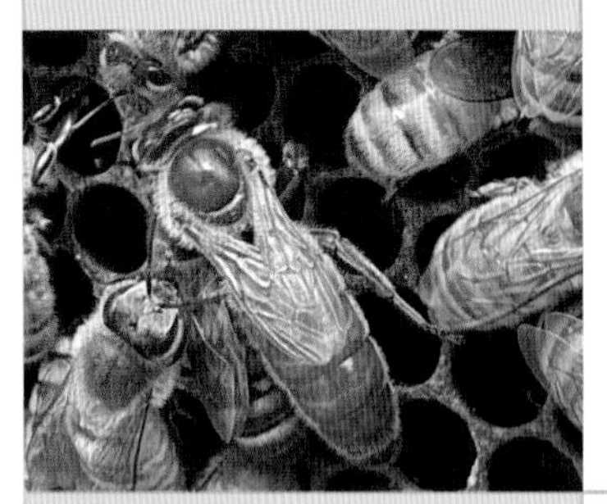

The most common and popular bee for urban beekeeping worldwide, this variety of honeybee is gentle, productive and tough. The queens will be either caramel coloured or with tiger stripes, and stand out a bit more against the workers. Italian queens create colonies that build up in numbers relatively gently, making swarm control a little easier. They very rarely swarm in the first year, and many beekeepers prefer them, changing to new queens annually. This breed is the best for rooftop beekeeping in my opinion, as swarm control is paramount in inner-urban areas.

Carniolan queens

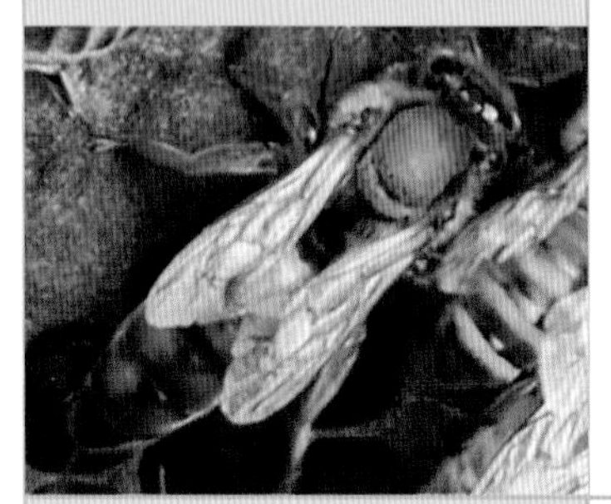

The Carniolan breed of honeybee originated in Slovenia and is sometimes called the grey bee, due to the fuzzy hairs that give the workers a grey shade. This is a very gentle variety and easy to work, but much faster to respond to environmental conditions, so you really have to be on your toes for swarm control! The queens are darker and striped, making them a little harder to spot, so they often have their backs painted with a dot of paint, like this one. This breed is best suited for community gardens, as they are gentle and will rapidly build for spring pollination.

Caucasian queens

This strain of bee has much darker, sometimes completely black, bees with longer tongues, and who work earlier and later in the day, even in cold conditions. Originating from Georgia in the Caucasus region, they are well adapted to the cold. They also produce much more propolis (a type of antibiotic bee glue) that is prized for cosmetics and medicinal treatments. The extra propolis is also purported to give them improved disease resistance. They are, however, a somewhat shorter-tempered bee in my experience, so may be better suited for rooftop apiaries with limited interaction with the public!

Ligurian queens

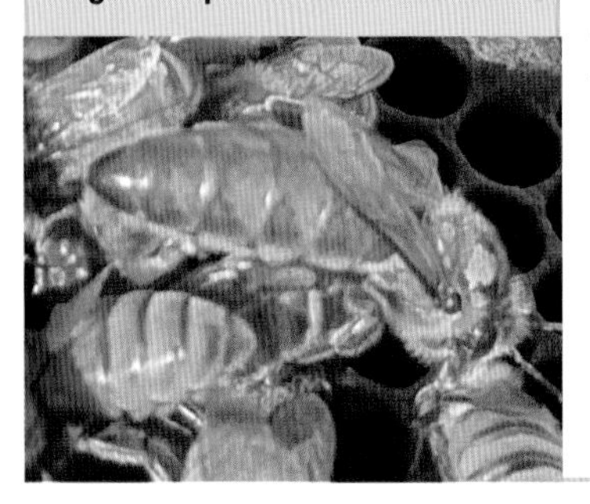

Produced on the bee sanctuary of Kangaroo Island, which was declared in 1885 to preserve the unique genetics of this strain from northern Italy, these queens are purported to be the last pure strain of honeybee left in the world. They are productive, gentle and produce a beautiful honey, but tend to struggle in colder winter environments. The apiarists of Kangaroo Island were badly impacted by the Black Summer fires of 2019/2020, but have been rebuilding their stocks of bees to produce queens once again.

Within these breeds, there are additional traits that the breeders may select for, and there will be times when you specifically want these. Whenever you see the term 'hygienic' in the title, it means the bees have been selected to obsessively clean their hive, which is particularly useful for fungal diseases like chalkbrood (see p. 150). Changing to a hygienic queen is one of the best treatments for this disease, and it does not affect their overall temperament. They are generally more expensive, though!

There is also an international push to create Varroa-specific hygiene traits, which include bees that aggressively attack mites clinging to adult bees and groom them off, and bees that can identify infected brood cells and clean them out. Still very much a work in progress, these genetics are beginning to be imported into Australia to allow us to prepare for possible Varroa incursions. This will be critical if the current incursion around Newcastle escapes, or if any future incursions get away.

HIVE REGISTRATION

Registration of hives is mandatory in all mainland Australian states and territories except for Tasmania, where it is voluntary (but highly recommended). For recreation and small-scale beekeepers, registration is either free or attracts a small fee, and carries a number of requirements. These include the requirement to clearly mark the registration number on the outside of all boxes and following the Australian Honeybee Code of Practice, which is mainly concerned with biosecurity.

The most important part of the hive registration process is your contact details. This is the real benefit to you from registration, as the biosecurity officers can get in touch to warn you of pests and diseases in your area. It has saved my apiaries through early inspection in a few cases, catching diseases before they can spread.

FOOD STANDARDS

There are myriad food standards requirements throughout Australia, and these are generally managed at the local government level. There are a few common essential elements that you must adhere to if you plan to sell your honey:

- **You must register as a food business, and this usually includes developing a**

Food safety always needs to come first.

food recall plan in case of contamination of a batch of honey (rare but possible, for instance if the bees pick up a pesticide).

- You must use an approved food space to extract and pack your honey. Be aware that turning your home into a registered food business will sometimes require you to take out business insurance, and will not be covered by your home and contents insurance.
- You need to have a registered business name (make it a catchy one; bee puns are allowed!).
- You need to create a honey label that lists the contents (which is usually just 'honey'), the weight of honey in the jar, a batch number, the nutritional information, and the label must be a food-safe adhesive label. It also must contain your name (or company name) and address. This cannot be a PO Box or website — be aware that this will probably mean your actual home address is on your jars.

When extracting for Parliament or one of my corporate clients, I have a registered business name and use their commercial kitchens to extract and pack the honey into jars. I also have labels professionally designed that comply with food labelling laws. Yes, it is fiddly and annoying paperwork to do all of this. No, it is not optional. Get a checklist, get registered and make sure you are legally compliant.

QUARANTINE AND BIOSECURITY

Australia is unique when it comes to quarantine laws. We are the last continent still largely free of the major bee pest, Varroa mite (see p. 154), and the sheer size of the country means we have multiple quarantine zones. Most countries just have one, and often don't police their borders too vigorously, which is why they have so many pests and we don't.

This means you need to be aware of quarantine zones and not try to send products to certain areas of the country. Western Australia, in particular, is free of all major pests and diseases of honeybees, and wants to keep it that way. Trying to send honey or wax to the west, or using non-WA beekeeping gear there, will most likely result in confiscation of the items or a hefty fine, or probably both! Western Australia produces some of the most beautiful honey in the world, which they can send to you anytime you desire some. Just don't try to send anything back!

It is essential that you understand what a notifiable disease is (see Chapter 9). Some diseases like chalkbrood are considered common enough that you don't need to tell anyone if you find them — just manage them. Notifiable diseases like American foulbrood are much more dangerous, and must be reported to the chief veterinary officer in your state or territory. Each jurisdiction has a list of notifiable pests and diseases, and you should get regular updates via your registration contact emails.

INSURANCE JUST IN CASE

Do you need insurance? Yes, you absolutely do! Just as most of us would never consider driving without insurance, you should not drive a public beekeeping operation without this backup! In the case of Parliament House and my corporate clients, being insured is mandatory. There are a few traps for new players, however, and you need to know about these when you transition from backyard to urban beekeeper.

Most bee clubs around the country offer some form of group insurance, and this is a good deal for the beekeepers and the insurers. The beekeepers usually get a group discount, and with many clubs having upwards of 500 members these days, the insurance companies get a larger pool of customers. The beekeepers who join a club to improve their skills are often the right sort of beekeepers — interested in continually improving, and receiving advice, training and support. The problem arises when you transition from a few hives to a business that is selling honey for profit, and running tours, courses and other events. This is almost never covered by standard beekeeper insurance and requires a specialist (and more expensive) package.

Common 'red lines' you need to let your insurer know about include when you add any of the following:

- **honey sales, especially from your house or the community garden**
- **receiving payment for swarm collection (even as little as $20 to cover fuel)**
- **public events or visits of any kind, including field days and training**
- **employing staff, even as occasional casuals**
- **accrediting your kitchen or shed as an approved food space (this usually requires business insurance)**
- **reviews after any major incidents.**

It is not a big deal, unless you don't talk to your insurer about the changed circumstances — then it becomes a very big deal indeed. In that case, you are essentially paying money for nothing; your insurance will be voided and it could even void your other insurance policies.

Have the conversation. Get specific insurance, with your business activities set out in writing. Yes, it will probably cost more money, perhaps even a lot more — you need to know this and include it in your prices!

RISK REGISTERS AND SAFE WORK METHOD STATEMENTS

People often ask me why I ended up as the Head Beekeeper for the Australian Parliament, when there was probably interest from lots of other people. The answer is to some degree being in the right place, being polite but persistent, and also the famous overconfidence of a middle-aged white guy! There was one other factor that was important: I am good at risk management, especially for environmental work like beekeeping.

It might seem boring (and it is) but a good risk-management plan is one of the most important factors in moving from 'I want to be an urban rooftop beekeeper' to actually getting approval to do it. It is also critical for when things go wrong, and it can prevent things from going from bad to worse.

The first thing to do, especially when being invited to keep bees in a new space, is to walk through the area and note down the tasks you will need to do in order to work bees. Then a

One of the access points for a rooftop apiary — lots of places to trip and fall!

Lifts are absolutely the best way to get hives to the roof!

bit of lateral thinking comes in — what are the hazards that could happen? We often focus on stings, but the reality is that slips, trips, falls and objects falling from heights are far more dangerous than any bee sting could ever be for most of us, and are much more common.

I was once asked to look at a rooftop space where the only access was via a two-storey-high vertical ladder with no fall-arrest hook-on points. This would have required winching up gear, including hives weighing over 50 kilograms, where any failure in the ropes would have had a potentially fatal outcome for myself and others. That was a very hard pass from me!

Next, you need to think how likely these hazards are to actually happen, and focus your efforts accordingly. A meteor strike on your hives is possible, but unless you are a superhero with alien enemies, probably not worth managing for (if you are, please call me — you sound awesome). On the other hand, tripping over happens all the time and carrying a heavy box of gear can definitely result in a twisted ankle or a broken bone. You need to actively manage these risks.

CORMAC FARRELL - HIVES AT 7 LONDON CCT & 4 NATIONAL PLACE, CANBERRA - RISK REGISTER

Risk No.	Date Opened	Objectives *What are we trying to achieve?*	Potential Risk *What could go wrong?*	Current Risk: Likelihood	Current Risk: Consequence	Current Risk: Risk Rating	Risk Mitigation Strategy *How do we plan to manage the risk?*	Residual Risk: Likelihood	Residual Risk: Consequence	Residual Risk: Risk Rating	Responsible Officer	Status	Date Closed
1.	HIVE MANAGEMENT												
1.1	11 Jan 2019	Hive health and disease management	• Disease or other infestation may impact colony strength and honey production • Disease and other infestation may spread to other nearby colonies.	3	D	MODERATE	• Beekeepers have adequate knowledge of potential pests and disease through information sharing and training • Beekeepers will conduct regular inspection and monitor hive health • Beekeepers will prepare maintenance schedule and identify back-up resources	1	E	LOW	CPB	Open	N/A
2.	OPERATIONAL												
2.2	11 Jan 2019	Minimising incident risk from smoker mishap	• Smoker mishap leads to a fire hazard	2	D	MODERATE	• Beekeepers will place the smoker in a metal bucket when not in use. • Beekeepers will take due care when lighting smoker to ensure that no naked flame is exposed. • Beekeepers will maintain a bucket of water. Ensure all the ash and fuel in the smoker is thoroughly soaked when finished. • Beekeepers will never use smoker on Fire Ban days	1	E	LOW	CPB	Open	

An example of one of my risk registers.

The final thing is to start designing your safe work methods — these are the actual controls you put in place to manage the risks. The key concept here is something called the hierarchy of controls. Not all risk management is created equal, and you need to design controls by working from most effective through to least effective measures. We can use the ever-popular risk of honeybee stings to illustrate:

Elimination **Physically remove the hazard.**	Keep native bees instead of honeybees. Provide a live video feed of bees from a remote site.
Substitution **Replace the hazard.**	Replace wild honeybee swarms with a more docile strain of bees.
Isolation **Keep the hazard away from people.**	Place honeybee hives in an area where people rarely go and that has controlled access (e.g. locked doors, swipe pass). Isolate the hives behind a glass window.
Engineering controls **Engineer out the hazard.**	Install screens to direct honeybee foragers up and away from where people are.
Administrative controls **Change how people interact with the hazard.**	Install signs to warn people to stay clear. Use sign-in sheets at the entrance to acknowledge the risk.
Personal protective equipment **Protect yourself and others with personal protective gear.**	Wear a bee suit when near the hives. Have spare suits, smokers and gloves available near the hives. Keep an EpiPen on site in a clearly marked first aid box.

This probably seems like a lot of paperwork and thinking, but it is worth it. You will be surprised how quickly you can work through a risk register, and how effective it is in identifying risks. Most people who have worked in construction management or environmental survey will have done these, so reach out and get help.

Now that you know the sorts of administration you need to have in place to be a safe and effective urban beekeeper, it's time to find a place to keep the bees!

4
A place to bee at home

Where to keep your apiary

'We have a place, you should come and keep bees here!' This is a phase that most beekeepers will hear on a semi-regular basis. Sometimes it works out, and it's awesome. Sometimes you run a mile in the opposite direction.

There are a few key things that make for a great apiary site in urban areas, and they are very different to where you would keep hives in a rural setting. There are some unique risks in the urban environment, but also some unique opportunities that make it very worthwhile. So what factors do you need to think of?

Most people start beekeeping as an add-on to their home garden, but large blocks with generous space for gardens are increasingly becoming a thing of the past. In response, community gardens started appearing, and these range from a few simple garden plots to sprawling complexes complete with their own food outlets and seed-saver networks. This is not a new thing. There was already a thriving community garden movement throughout Australia, but the Covid lockdowns supercharged it as community gardens became an outdoor space where you could safely do something as a community that was both productive and interesting.

The Parliament gardens are almost the perfect place to keep a public apiary. (Credit: 5 Foot Photography)

You do need a certain amount of room for a honeybee hive for safety, but anyone can have a native bee hotel, and every community garden should. Bees are one of the most important pollinators, creating larger crops, and tastier and more nutritious food. They also allow the garden to act as a mini-refuge for native species, and increasingly these showcase organic gardening methods that teach people to live without spraying everything in sight.

For the urban beekeeper, community gardens are also perfect for outreach into the wider community. They are by their nature easy to access, and bring together lots of different folk from all walks of life.

A decent apiary has some common elements that will really help you to enjoy the experience.

FOLLOWING THE SUN: POSITION IS EVERYTHING!

I love waking up to the beautiful morning sunshine, and my bees are no different. For a honeybee hive the ideal position is an area that gets morning sun and afternoon shade year-round, with an emphasis on sun in winter and decent dappled shade in summer. The latter is particularly important as heatwaves become more prevalent around the globe, and keeping the blast of heat off the hives can be the difference between life and death for the colony under these conditions.

Trees provide the best shade, closely followed by buildings where, ideally, native bee hotels or honeybee hives can be sheltered from the wind while still catching the morning rays.

WATER FOR WILDLIFE: BEES AND MORE

You might think that I would put flowers ahead of water in terms of importance for bees, but flowers are a bit of a given in a community garden. Water, on the other hand, is one of the most overlooked parts to building a biodiverse garden. It is also one of the easiest things to fix. From a basic dish filled with pebbles to a complex solar-powered pond system, adding water to your garden will help the birds and the bees, as well as adding beauty and creating a favourable microclimate.

Water features are not just a 'nice to have'; they are essential to keeping many bees alive, especially honeybees. With more frequent extreme weather events like heatwaves on the way, they are also an essential adaptation to cool our urban environments and protect our wildlife. We saw this firsthand on the east coast in the aftermath of the Black Summer bushfires, with people putting out bowls of water for birds and other animals fleeing the destruction and often taking refuge in urban areas. While it was heartbreaking to see exhausted-looking birds with burnt feathers resting in these temporary ponds, for many of them it would have been the difference between life and death. We don't have to wait for a firestorm to start making our urban areas better for wildlife — we can all start right now, and community gardens are one of the best places to showcase this.

Bowl of pebbles or corks

This is the simplest watering system for wildlife: just a deep bowl filled with enough pebbles so that bees and other small insects don't drown and can drink while standing on

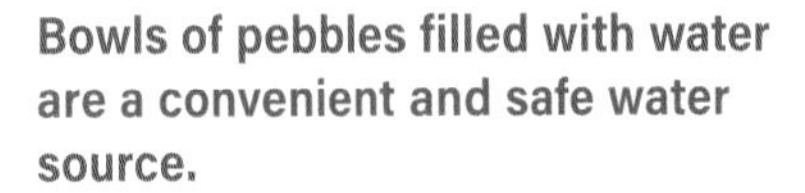

Bowls of pebbles filled with water are a convenient and safe water source.

Pond systems can be a bit of work, but your bees (and frogs and birds) will love you for it.

something. For extra sparkle, I have seen some beautifully glazed bowls filled with marbles, which creates a spectacular look when the sunlight shines through them.

Birdbaths

I have several birdbaths in my garden, and I love watching the wildlife that comes to visit. The best thing about these is that they usually have a shallow lip that allows bees, lizards and other small critters to drink safely, while having a deeper centre that birds can bathe in. They also come in beautiful and artistic shapes. My favourite birdbath in the garden is a statue of a frog holding a lily pad above its head.

Pond systems

Even a single pond in the garden can transform the feel of the area and creates a cool microclimate. I have a series of ponds linked by pebble-filled creek beds and driven by solar pumps. Not only do they guarantee that my hives have water, they make for a beautiful garden feature that lets me grow a wider range of plants. It took me a weekend to dig out and set up, but the ponds have fish and plants that keep the water clear, and the pumps just need their filters checked every now and then. Highly recommended, and worth the effort.

This urban wetland controls stormwater flooding, and is an oasis of green in the middle of a busy suburb.

Wetlands

These are the big daddy of water sources, and they are increasingly appearing in urban areas to control stormwater flooding and improve water quality, so these may be close to many community garden spaces. Wetlands can be natural, but increasingly they consist of engineered ponds where stormwater is diverted, with planted wetland species at the margin being used instead of chlorine to keep the water clear and clean. They are stunning, but also expensive and take some skill to create and maintain. That said, having the margins planted with a mixture of aquatic plants makes for a spectacular oasis for a wide range of urban (and human) wildlife. Plus, they are often beautiful in their own right, and water our bees!

How much water?

If you keep honeybees then your water sources should have a minimum capacity of 2 litres per hive — this is how much a hive can go through in a day, and ideally you should have much more than this to ensure they never run out in hot weather.

One important consideration (and a legal requirement in some areas with significant mosquito-borne disease) is to prevent the build-up of mosquito larvae in the water. Adding fish to ponds (particularly small native fish) and periodically tipping the water out of birdbaths before refilling is an effective way to stop your water sources from becoming breeding sites.

GOING NATIVE: BEE HOTELS

It is not realistic or even desirable for everyone to keep honeybees. Sure, they are adorable, produce food and carry out pollination, but they are also interlopers to the Australian environment. Even where they are native, it is common for limits to be placed on how many hives can be put in a given area to stop overcrowding (see p. 27). Plus, there are the stings and the responsibility of caring for a large, sometimes cantankerous colony. What if we could get that feel-good 'save the bees' vibe without all the work, mess and funny suits? In that case, we have a deal for you — native bees to the rescue. And you get to rescue native bees. Convenient!

Our cities are about as far as we can get from natural environments, which is all the more reason to make them a bit more friendly for native species. Two things are often in short supply for native bees in cities: native flowers and nesting sites. Flowers we have already dealt with, but just like the rest of us, the real estate market for bees in the city can be pretty dire.

The thing I love about native bees is that providing nests for them is really low maintenance and rewards tucking nesting areas into lots of small spaces throughout your community garden. There are literally thousands of bee species throughout Australia, and worldwide, tens of thousands. Almost everywhere has dozens of different species and keen gardeners start to treat it almost like a challenge to collect them all.

Some of the basic types of native bee hotels that are popular with gardeners and the bees include the following.

Wood blocks

These are the simplest and give you the highest chance of attracting some colourful, fascinating residents. Simply drill deep blocks of wood with lots of holes, drilling with the grain to get a nice smooth finish on the interior. I use leftover bits of firewood and go through my drill set, peppering the wood with lots of different sized holes. The deeper the holes drilled into block the better. Native bees tend to lay male bees towards the entrance and females deeper in the cavity. If you only drill shallow holes you are only

Community garden bee hotel — simple and effective.

The filled holes mean that this bee hotel is proving popular.

Bundles of bamboo or sunflower stems make up this native bee hotel.

accommodating for males. You also run the risk of having the entire nest parasitised.

Another critical factor that is often overlooked is how smooth the holes are. In nature, these bees use holes made by wood borers, which cut a glassy-smooth path through the trees they feast on. Holes drilled with a cheap drill bit can leave rough edges and splinters, which will shred the delicate wings of the bees as they enter and exit. Make sure you use high-quality, sharp drill bits, and that you haven't left any splinters or sharp edges at the entrance that could catch the bees out. Drill with the grain rather than against it.

Bamboo

Bundles of bamboo stems attract a wide variety of native bees and wasps, and they are all beneficial for the garden. The most common will be small native bees from three genera that are broadly referred to as Reed bees due to their preferred nesting material. These are dead simple to make. Just cut the bamboo (or even plant stems) to around 20-centimetre lengths, gather up fifteen to 20 stems in a loose bundle, tie the bundle in a few places and then hang it under tree branches or eaves. Don't stop at one — put up as many as you have room for, and you should see the ends of the bundles being excavated and then capped off as the bees move in.

Mud bricks

These attract the ground-dwelling bees like Blue Banded bees and Teddy Bear bees. Unlike the other natives that seem to enjoy smaller nests spread through the garden, these bees really like apartment living, with a stack of mud bricks bringing the best results. I have used bits of waste pipe cut into lengths and stacked up, and also found success with stacks of Besser blocks filled with mud. Once again this is easy and fun. Fill the blocks or pipe sections with mud, and then stack them somewhere sunny but sheltered where you can get about ten bricks stacked up. The best mud in my experience is red clay, and I know that some people mix some sand into it to make it easier for the bees to dig out.

It might take a while for the bees to move in to these 'bee walls', but you can speed this up if you place them near to where the food is — Blue Banded bees love tomatoes and Strap Lily (*Dianella* spp.). If you want to get fancy, you can hang a length of string off a stick out the front — the males will sleep by hanging onto this with their front mandibles while they wait for the ladies to emerge from the holes.

Maintenance of native bee hotels

Native bee hotels are definitely lower maintenance than their honeybee counterparts, but they are not 'no maintenance' either. Over time, the nesting materials will build up disease

The filled holes in these mud tubes mean lots of Blue Banded bees in spring!

A mud brick native bee hotel in a Canberra park, surrounded by native plantings.

Bamboo straws filled with resin bee brood.

A large native bee hotel installed at a community display garden.

spores and waste, and they can also start to attract predators. In nature, this is less of a problem; the natural processes of a forest spread the nest sites out and also continually renew them. In an urban environment, these processes don't work as well, so we need to help out.

I like to cycle my wood blocks, straws and mud bricks every few years. When changing over, I try to put the new nesting hotels within a few metres of the old ones, and then take away the old material once I can see that all the baby bees have hatched out. With some designs like wood blocks and bamboo straws, you can also clean the holes out with a pipe-cleaner dipped in soapy water.

There are a lot of commercially made bee hotels available in garden supply stores, and even supermarkets, but they are often mass produced as a 'feel good' gimmick. These are controversial in the native bee community. While they do raise awareness about native pollinators, they are often quite poorly made. Shallow holes, short straws and rough edges make some of these barely viable as nesting habitat.

There are also some beautifully made bee hotels. My favourites are made by the Flow Hive folks and there are others made by specialist native bee keepers. Some of these hotels even have viewing windows that let you look in on the baby bees. These are more expensive than the mass-produced versions, but still relatively cheap. Most importantly, these higher-quality hotels will give your garden pollinators the best chance of surviving and thriving.

For those who want to make their own, native bee researcher Dr Kit Prendergast has created one of the best resources covering design of the common hotel types, where to place them and even garden species lists. Titled 'Creating a Haven for Native Bees', it is available for order on her website and is one of my 'go to' resources when building bee hotels.

Community gardens allow you to recruit friends to help!

Sweet success: Adding honeybees

There are lots of great reasons to add honeybees to a community space and, to many people's surprise, honey is a little way down the list. For many community gardens, the real reason to include honeybees is the same as for agriculture: pollination. Plants can't move and must use either wind or a pollinator like a bee to spread genes between plants. Native bees do some of this, but for many fruit trees, berries and vegetables, honeybees are their natural pollinator, and are ideally suited to the job. You still get the honey, of course!

With my local bee club, our main training apiary sits within a community garden and horticulture training complex called 'City Farm', which teaches urban sustainability. It is the perfect setting to teach about integrating bees into gardens and horticulture.

Staying clear of the stings

Honeybees are happy to live in almost any situation; much like us, they are incredibly adaptable and clever at finding places to live. However, for safety, and in some areas legal compliance, there is a minimum space that you should have around the hive, as well as some practical considerations to let you inspect and enjoy your bees. You need to have a minimum of at least 8 to 10 metres of clear space around hives, which gives the bees plenty of room to come and go and, importantly, means that the hives are not too close to garden beds or paths where people will be walking. When honeybees emerge from the hive, they will immediately head in the direction of the flowers they want to visit, and will climb to their cruising altitude that puts them above obstacles. As a result, the main area of traffic where you can get in their way is about 3 to 4 metres from the hive. If there is something like a fence or a trellis a few metres from the hive they will fly up to avoid this and then keep going up, so this can be a good way to get them to climb up and away from where people are.

STANDING ROOM ONLY

Room for you to stand during inspections is also important — you don't want to be standing in the way of the foragers as you inspect, as this will annoy them. You want at least a few metres of good space at the back of the hive so that you can lift boxes off and place them somewhere close by, and you really want this space to be level. I once established a hive on a sloping bank, cutting a small space to level the hive but leaving a sloping space for me to stand — never again! Falling over while holding a frame instantly covers you with confused, and then angry, bees. Keep the space level, and remove any trip hazards, ropes, bricks and other things that are not required.

The other nice thing is a space off to the side of the main traffic area where you can watch the bees work. This is one of the less-publicised delights of beekeeping; it is lovely to just watch the bees come and go and is a very practical way to put you in tune with what might be going on inside the hive (see more on this later).

Room for selfies is also a critical consideration. (Credit: 5 Foot Photography)

Lighting the smoker in a clear area with water close at hand.

Setting up for inspections

The other important thing to consider is your staging area where you will organise your gear, put on your suit and light your smoker. This is almost as important as the hive site itself, as it lets you have extra gear like nucleus hives (see p. 99) readily at hand in case you need them. This really needs to be a bit away from the hive, at least 10 metres or so, to allow you to shake off any stragglers or enthusiastic/cranky defenders if you happen to annoy them. A good hard-stand area to light your smoker is also critical. You definitely don't want to be trying to light your smoker in among grass or anything else that could be flammable, especially in summer! Remember that you should always have a bucket of water handy just in case, and this needs somewhere to sit as well.

I personally like to set up next to an outdoor table. I have a seat to help with getting the suit on, a place to lay out spare tools and extra gear, and a hard paved surface to light the smoker on.

REDUCING THE TOXIC COCKTAIL

The policies and politics around pesticide use have been a constant focus of the environmental movement since Rachel Carson's seminal work, *Silent Spring*, exposed the impacts of indiscriminate pesticide use. Many of the issues she identified remain with us today, and for many of the same reasons. We absolutely need pest control for both our food production and for healthy gardens, and pesticides are an effective tool; however, this comes at a cost. This cost increases sharply when pesticides are used excessively, indiscriminately, or without thinking of the potential risks.

Chemical companies are usually presented as moustache-twirling villains in any discussion around pesticides, and at times they do seem to act the part, but mostly they are businesses trying to make money. One of their biggest moneymakers has also been among the most damaging for bees: systemic pesticides. This is a class of pesticides that works in a fundamentally different way to the pesticides people are used to using. Rather than trying to spray the toxin on the bugs to protect the plant, systemic pesticides become part of the plant, essentially turning it

Credit where it is due — many garden stores have replaced systemic insecticides with more bee-friendly non-residual products.

into a toxic deathtrap for pests. These chemicals even stay resident in the soil, incorporating themselves into successive crops of plants, sometimes for years.

Systemic pesticides have proven to be great for business, in that farmers and home gardeners will buy them whether they need them or not; they are incorporated into seed coatings, which then follow the plant into the ground. Companies have made high profits and are able to predict more accurately how much will be used in each year, reducing production costs. They are also very specific to insects and are generally safer for human health.

This suburban hive exhibited all the symptoms of pesticide poisoning.

In the process, these companies have created a product that pervades the environment long after its original use is forgotten. This is particularly bad when sold to home gardeners with little training or understanding of chemical use. How many gardeners realize that the spray they used for those bugs two years ago is still resident (and lethal) in the soil now?

So we see the problem, but what is the solution? The answer from the chemical makers is typically the classic lobbyist scare tactic: if you don't let us sell systemics then people will have to use the old chemicals, which are even worse. However, there are many alternatives to systemic chemicals and not all of them are in the realm of organic farming. Many conventional farms are switching over to less costly and more environmentally friendly approaches, and city managers should be taking note.

The most basic question that we need to ask is, do we even have a problem that needs a spray? People get really caught up in having perfect plants, and this attitude flows all the way through from home gardeners through to park managers, where a chewed leaf is seen as a sign of neglect. However, it is also a sign that a caterpillar was living there, and that it is trying to turn into a butterfly. Or maybe a songbird caught the caterpillar and is feeding its young with it. That ragged, chewed leaf is not a waste — it was food for something. Perhaps something wonderful.

Throughout the world there has been a widespread loss of insect biomass, often given

The Old Parliament House rose gardens thrive despite minimal sprays being used.

the dramatic moniker 'insect Armageddon' in the media, but this is a real and very scary problem. There has been a very significant and measurable decline in insects, around 80 per cent according to a long-term German study.[1] Most worryingly, the decline persisted even inside nature reserves. At the same time there has been a decline in songbirds throughout Europe, and this has been matched by declines in Australian species such as the plucky Willie Wagtail. Our landscapes are becoming silent, and drab.

The simple fact is that insects eat stuff to stay alive, so we have to tolerate them taking a few nibbles here and there. The question is one of balance. Tolerating some holes in a leaf means that we are delighted by the colours of a butterfly a few weeks later, and our bees stay safe from sprays. That seems like a reasonable trade-off to me. Of course, something that is in balance can also get out of balance, so we need to learn (in some cases re-learn) to recognize when the pests are getting too far ahead of the natural controls, and what to do if that happens.

MINIATURE MONSTERS: YOUR GARDEN FRIENDS

The Australian Parliament is showing the way in managing gardens in a minimal-impact style through the use of integrated pest management (IPM for short). This is all about understanding the biology of the species you are trying to control, and then using the minimum effective intervention to achieve what you want.

The first question to ask is, what do you want? Do you really need picture-perfect plants with no holes in the leaves? If you are trying to win the flower show, this might be a consideration, but most gardens don't need such aesthetic perfection and keeping all of your garden completely 'bug' free is likely to come at a significant cost. It also puts you on a treadmill of continually buying sprays: in eliminating the bugs that annoy you, the bugs that eat them also die off. As a result, all responsibility for pest-control then falls on you as the gardener, meaning you're endlessly buying sprays and then spending hours applying them. Great if you are the person selling the sprays, but bad for the bank balance and the free time of the gardener.

The alternative is to cultivate your own posse of miniature thugs to beat up the bugs for you. In addition to their voracious appetite for your garden pests, they also add colour, movement and interest to your backyard patch. Simply keeping my cats indoors all the time, mulching and creating a water source attracted myriad pest controllers to my home garden. These ranged from frogs, several different species of lizard and insect predators like praying mantis. All have different preferences for the pests that they eat, but equally they all need the same basic things — something to eat, somewhere to live, and safety from chemicals, domestic cats and other predators.

Huntsman spiders are great pest control for the garden — at the cost of the occasional bee.

This beautiful gecko was happily living under the lid of my garden hive.

Sprays that don't stick around

I work on a wonderful organic farm managing the pollination services, and one of the major misunderstandings of such farms is thinking that no pesticides are used at all. This is not correct — they are, but organic-certified pesticides often have a few things in common that reduce their harm. Firstly, they do not have a long residence time in the environment, breaking down quickly (usually within 2 to 6 hours). While this means they have to be applied quickly to be effective, it also means that a spray applied in the evening after my honeybees have gone to bed will break down by the time the sun rises and will not kill my bees or contaminate the honey. Secondly, these pesticides do not become part of the plant, meaning that, once they have broken down, they do not pop up again in the soil, nectar or pollen. There is a range of products available and they are often quite specific to the pest or disease that they target, but non-residuals are normally clearly marked as such, ideally with how long they take to break down in the environment.

Of course, this only works for honeybees; many native bees roost at night out in the open and could still be hit. There are a few general approaches that work well with non-residual pesticides to further reduce the risk:

Applying in the evening: Almost all bee species go to bed in some form at night, so you will have a pretty decent window of time where they are not foraging on flowers.

These flower buds are several weeks from opening and are not attractive to bees.

Application method: Do you need to spray the whole plant or just the parts that are being attacked? If you are targeting aphids or caterpillars, they are rarely on the flowers themselves, but on the leaves and stems.

Timing of sprays: How long are the flowers going to be on the plant? Some plants have a distinct flowering period, and if it looks like the pests might hang on until after the plant finished flowering, then you can treat the pests then. This particularly applies to aphids and other sap-sucking pests, which love to target the young shoots and flower buds before they open.

WEED CONTROL, BEE-STYLE

Burning

The indiscriminate use of herbicides is as much of a problem as the use of insecticides, as it kills a wide range of non-target plant species, many of which are valuable food. At a wider landscape level, herbicide creates a lot of problems with runoff and herbicide resistance. Resistance is a progressive problem where the weeds become increasingly tougher to kill, requiring more chemicals to be used for the same effect. The costs and wider impact keep going up and up, and in extreme cases chemical controls can become unworkable.

The good news is that there is a range of effective alternatives to chemicals, and these work at both a home garden and city scale. One of the most fun is a weed burner, which is basically a hand-held blowtorch powered by a gas cylinder. Obviously fire is dangerous and this has to be used carefully, and never during high fire danger days, but it is an extremely effective way to manage weeds. Rather than trying to burn the weeds up entirely, the gas burner superheats the water in the stem and base of the plants, killing them in a few seconds. It is an effective way to manage plants emerging between pavers, and we use it in the training apiary to keep the pads clean of excess grasses (the hives sit on gravel that rapidly heats up, making this even more effective).

Mulching controls weeds, conserves moisture and builds healthy soil.

Weed burners are fine for community gardens and small spaces (where there are fewer combustibles) but what about doing this on a city scale? Turns out there is a great solution: steam. Industrial steam wands are truck-based units that can generate large amounts of superheated steam. This has been proven to be comparable in effectiveness to glyphosate, and is commonly used in the greenhouse and nursery industries to sterilize soil before planting. The biggest advantages are that the residual material is just water, and that weeds have a tough time adapting to high temperatures. Some types of steam rigs also double as cleaners for graffiti, making this a versatile piece of equipment.

Dandelions will almost always find a way to emerge through mulch, but that is not always a bad thing!

How could you not love this face?

Mulching

Applying mulch to soil is one of the oldest and most effective soil conservation techniques, and is something that every gardener should be doing, whether they are at home or on a city scale. Mulch can be basically anything that covers the soil, preventing weed seeds from germinating and reducing moisture loss. Common mulches include wood chips, straw and leaf litter. In fire-prone areas close to buildings, pebbles and stone chips are also highly effective. In my home city of Canberra a significant portion of the mulch comes from the removal of unsafe or damaged trees, which are chipped and stockpiled, creating a cycle of re-use.

One of my beautiful rooftop apiaries.

Mulching doesn't just prevent weeds from emerging; it encourages soil life and reduces evaporation. Mulch is also easy to apply — just spread material so that there is an even covering 4 to 8 centimetres deep across the top of exposed soil, and you are basically done. Organic mulches gradually break down into the top layers of soil, and these will add organic material, further increasing water-holding capacity. Just remember that not all weeds are really weeds; leave a little room for flowers like Dandelion in among the neat garden beds!

A good coarse mulch is not just great for weed control; it is also pretty fantastic for insect control as well! The mulch creates its own little ecosystem, complete with lots of bugs, which then attracts lizards, frogs, small birds and spiders. This creates a pool of predators that can respond to any build-up of pest insects, and you will quickly see a surprising variety of garden friendlies turning up. My absolute favourite are the gecko lizards — I'm not sure if that's because of their frog-like faces or the fact that they keep the ants away from my hives. I even modified the design for my Warre hives to give the geckos a space to live in the roof, and the bees seem to be perfectly happy to have lizard flatmates.

Just as it is important to have areas with good levels of mulch, for optimum bee-friendly gardening you should also have areas of bare ground with no mulch. Quite a few of our precious native bees nest in the ground, including some of the early-emerging Golden Banded bee (*Tricholletes* sp.), which are among the most important early-spring pollinators where I live. Everything in moderation, including mulching, is the key to a diverse, vibrant garden.

GETTING AIR: BEEKEEPING ON OFFICE BALCONIES AND ROOFTOPS

One of the great joys of city beekeeping is to finish an inspection at the top of a tall building, then quietly sit by the hives as the sun sets on a warm summer evening, looking out over your city. Tall buildings provide a spectacular and productive way to keep bees and produce a surprising amount of quality honey. They are also one of the most challenging situations to work in, requiring careful management of a wide range of hazards.

Office buzz: The Aurecon office bees

Originally starting as a single hive on the first-floor balcony, the bees of the Aurecon Canberra office eventually took on a life of their own, inspiring food and artworks. Being primarily an engineering and project management company, the risk management and safety systems also laid the foundations for how we could safely manage a whole range of public apiaries. We originally kept the bees for fun, situated near a large window so we could watch them work while we worked (they worked harder, but not by much).

In many ways this was the perfect start to city beekeeping — we never used the balcony that they were on, and the large windows meant staff could watch inspections up close, seeing the inside of the hives without having to suit up.

A single hive eventually grew into two hives, pollinating and producing honey from the street trees and gardens around the Canberra office. The landscape in the centre of Canberra proved to be surprisingly productive, and it was soon time to start harvesting from the hives. A fun internal brainstorming session and online poll settled on the brand name for our office honey, and this started to become a highly sought-after commodity. These days lots of businesses have their own honey, but back then it was relatively rare to see and certainly helped break the stereotype of buttoned-up engineers. The honey turned out to be fantastic, but the real highlight was bringing the whole office together for the harvests, with everyone bringing their families in to watch us take the honey out of the hives and spin it out of the frames, then pour it into jars. Not only did it bring the office together, it was an absolute hit with clients, and helped put our little office on the map.

However, the absolute highlight wasn't around the honey, but around art. About a year after

Office honeybees — our most dedicated workforce! (Credit: Kris Arnold)

Our office honey, the perfect client gift. (Credit: Kris Arnold)

This beautiful Indigenous beehive is now on permanent display.

we started, the company began to commission Indigenous art (and other events) as part of our reconciliation plan. Our office contribution was to have a local Indigenous artist (the wonderful Krystal Hurst) paint one of our beehives, creating a stunning pierce of functional art. After a few seasons the painted hive was retired and brought inside to become part of the art collection in the office foyer. The beautiful colours and story of the hive were shared around the world, and the hive is now used as a showcase of artwork at UN World Bee Day events. It was one of the first major commissions for Krystal, and she went on to become an award-winning artist, creating jewellery and prints celebrating her Woromi heritage, and is about to launch her own fashion line. It remains one of the most beautiful and moving experiences that I have had in beekeeping. Work with artists whenever you can — like bees, they will make you see the world in a different way.

We had another coup when it came time to name the company's new intranet — out of all the offices across the world, our beehives inspired the name 'Hive', symbolizing working together as a team for a common purpose. Not too bad for a little regional office!

Eventually we had to move offices and were unable to take the bees with us, but that in itself created an opportunity that turned into the Parliament apiary, and with Aurecon becoming a co-sponsor for the Parliament bees. Proof that an innovative mindset and seizing opportunities can take you almost anywhere!

Thinking through the risks of each site and having a plan that you can implement and stick to is a critical part of successful rooftop beekeeping. It starts with a careful inventory of risks and systems that keep you and your bees safe.

Having a plan

Tall buildings have a unique set of hazards, and while they are a wonderful place to keep bees, they really aren't a good environment in which to wing it. Building owners will also almost always want to see a well-developed risk-management plan that includes safe workflows and a risk register. Yes, this is paperwork, but if they don't care about this it should ring some alarm bells for your safety; if they don't have a good system to manage risk, how do you know if you are safe on their site?

So, have a plan and be prepared to answer a lot of questions and adapt. This is my rundown of the main risks and a summary of the most common options to deal with the challenges that are likely to come up.

Getting up there – checking safe access

This is the cool part of tall building beekeeping — dodging your way through plant rooms, up hidden stairs and into areas that nobody ever sees. But this also presents the first real hurdle to establishing your apiary. You will be carrying all your gear in via this route, and boxes of honey out. There is also the chance that you might have to remove hives for building maintenance, taking live bees through the building.

Never agree to take hives into a building until you have sorted access, ideally while walking through it. In an ideal world you should eliminate stairs from the workflow as much as possible. Most modern buildings have a lift that goes to the plant room, and this opens onto the roof with minimal level changes, which is what you want. If you are unlucky, you might have to negotiate ladders or stairs, which can quickly become impractical when you are lugging a 20-kilogram box of honey. Any slip, trip or fall at this point could be disastrous for you and the bees, so think this through carefully, and walk it through with the building managers. However, complex access does have one advantage in that it almost entirely eliminates theft and vandalism.

When I was moving our original workplace apiary, the access became a critical factor. The building we were moving to had no safe roof access, a significant design flaw. The only way to access the roof was via a narrow shaft with a two-storey vertical ladder with no fall-arrest

These construction weights secure my rooftop Warre hives.

Rooftop hives strapped to a safety rail.

Rooftop apiaries are often accessed via the machine room on the roof.

devices. There was no way I was going anywhere near that, so the rooftop site was immediately excluded as a viable apiary. We had to have another plan to keep the hives, and that plan eventually became the Parliament House apiary.

Wind — your main enemy!

As a general rule, the higher you go the stronger the wind gets. This creates the single biggest challenge with rooftop beekeeping in cities: high-velocity winds. If you do not prepare for this properly it can be a disaster. Winds can easily top 100 kilometres per hour on the tops of tall buildings, tossing hives around, tipping them over and even entirely smashing them. This is a disaster for both bees and beekeeper, as the colony can die from exposure and will be extremely pissed off as you try to put the boxes back together.

You cannot rely on the weight of the hives alone, as high winds are still strong enough to toss the hives around, especially if the hive becomes lighter after harvest or a hard season (with less honey stores to provide weight). The hives really need to be securely lashed down to something solid. Railings, anchor points and support columns are your friends here, as they are securely anchored and are designed to carry loads.

Lashing straps like the ones used to keep surfboards on car roof racks are cheap, easily available and designed to secure loads against wind resistance, so don't skimp on these. Remember to not just have the hives lashed down to the ground, but to a vertical surface as well — this gives good resistance to buffeting winds that can toss and tip a hive.

It is worth looking up your local weather service as. The Bureau of Meteorology provides fantastic information at your fingertips. One really useful set of data that I use all the time in bushfire planning is something called a wind rosette. These are calculated for each weather station and provide an analysis of where the prevailing winds come from, but also show which direction the strongest winds come from. This is definitely worth checking out as you can place your apiary so that the building provides a shield against the strongest winds, meaning you only have to secure against eddies and gusts.

Sunshine – good for morning, bad for afternoons

Sunshine is great for bees. It wakes them up in the morning, signalling the foragers to start work, and it keeps the colony warm and dry. The exposed nature of many rooftop sites means that you need to pay pretty careful attention to sunshine, not just the total amount but also *when* the sunshine occurs. If I had to choose an ideal site it would have good morning sun and then deep afternoon shade.

Brilliant morning sunshine on this rooftop apiary.

One of the nice things about tall buildings is that you will often have a good choice of aspect for placing your hives, with the upper roof often having plant and equipment rooms, such as elevator machine rooms. You can really use these to your advantage, as they are a natural shade sail for the hives. Pick a north-easterly facing site if you can and angle the hives' entrances to face the morning sun, which will stream in and get the colony up and about early. The next priority is managing afternoon heat; hive roofs are very exposed and you don't want the hives baking in the summer sun. Try to pick somewhere that has good afternoon shade and is ideally near a tap so you can set up a watering station for the bees.

Sheltered site with morning sun and afternoon shade from the building.

Old light fitting recycled into a pebble bowl waterer.

Water — an essential ingredient

Water is an often overlooked (but essential) resource for honeybees, and rooftop hives are particularly vulnerable as they rarely have a water source close by. As noted previously, a hive drinks up to 2 litres per day in hot weather, and they need this for a range of things. The most critical is for temperature control in the brood chamber — under heatwave conditions the colony will stop foraging and collect water. They spit the water onto the outside of the hive and fan their wings in unison, creating a form of evaporative cooling. It is a classic indicator of heat stress to see the colony leave the inside of the hive and cluster around the front, making the beehive look like it has a beard. Perhaps unsurprisingly, this is known as bearding, and should immediately signal to you to check that they have enough water. Even without a heatwave bees still need to drink, so water supplies are essential.

A simple drinking dish such as a wide bucket filled

with corks or pebbles is easy to set up and ideal as a watering point for your bees. The key thing is to make sure the bees have something to stand on so that they don't drown. As with everything else on the rooftop, you need to guard against high winds. I fill my water dishes with pebbles, as this way the dish can't blow away and the pebbles also provide a platform for the bees to stand on as they drink.

Most rooftops have an external tap to help with cleaning, and you should place the watering point within easy reach of this but out of traffic areas where it could be a trip hazard. Don't worry if it's not that visible to you; bees can smell out water and will find it.

BEE-SCAPING YOUR COMMUNITY GARDEN

Helping bee populations is the classic case of thinking globally and acting locally, and it doesn't get much more local than your home garden. Obviously, not everyone has the privilege of a house and garden, and this has seen an explosion in shared spaces like community gardens, nature-strip plantings and even urban rooftops. There are so many things that you can achieve in even a small space, and almost all of these add to the beauty and utility of our cities.

They are just good friends. (Credit: Dr Kit Prendergast)

Start with the soil

One of the best general-purpose gardening sayings is, 'Grow good soil and the plants will grow themselves', and this holds true for almost all types of gardening. Healthy soil is fascinating to look at; far from being mere mineral dirt, it is a thriving community of plants, fungi, algae, microorganisms and critters. Hidden underground, this ecological community is usually invisible

Garden soil enriched with compost and alive with worms.

to us, but is the basis for much of the life on Earth. Soil is also the critical nesting resource for a whole cohort of beautiful native bees.

Here are some basic points to get you started creating vibrant soil that will support your bee paradise.

Organic matter

You need to feed soil in the same way as almost any other organism: with a wide variety of sources. Composts, natural manure, green manure (plants that are cut and dug in) as well as organic mulches all add organics. In addition to creating nutrients in the soil for plants, these also feed worms and other critters, which burrow through to aerate the soil. Soils with a higher organic content also retain water better, while still allowing drainage and aeration.

Structure

Ideally you want a light, crumbly texture to the soil that is a bit like chocolate cake. If you have to use a spade to dig into the soil, it is a sign the soil is too compacted. The best thing is that you generally don't need to do much. Just feed the soil organic matter and keep a light cover of mulch, and the tunnelling of earthworms, beetles and other critters will do the rest.

Bare ground

One of the quirks of bee-scaping is that you want to leave some areas of bare ground (space permitting), as this is key habitat for ground-nesting native bees. You don't need much, just some patches, ideally with some morning sun. Mulching absolutely everything can deprive these species of nesting habitat, and while you can replace this with bee hotels, small patches of natural soil are best. Check any bare patches for mounds of soil or telltale holes burrowed in before you spread mulch, just in case. This is one case where a bit of compaction is okay — the bees need firm soil to keep their home stable.

There is so much more to building soil; I have some colleagues who have spent much of their careers working on this fascinating part of our planet. Composting alone could literally fill several books, but the above will get you started on the basic components.

Water resources

As I have mentioned before, water is a critical resource for all bees, but honeybees in particular, with their large colonies filled with delicate wax sheets, are particularly vulnerable to extreme heat. In a major heatwave, the interior of the hive can become hot enough that the wax melts and slumps, suffocating the colony. To ward off this, honeybees have developed an ingenious response: evaporative cooling! As the temperature climbs, the foragers will start collecting water, spitting this on the outside of the hive, and then fan their wings in unison, evaporating the water and cooling the hive. It is incredible to watch, and when the colony is carting water in unison they can go through several litres per day.

Providing water can come in many forms — see pp. 63–64 earlier in this chapter to remind yourself. Interestingly, the one thing the water doesn't have to be is clean; bees seem to prefer muddy water for some reason, possibly to gain some minerals or salts. You also don't need to leave it out in the open; their incredible sense of smell will allow them to find water.

5.

Planting for bees

Changing the urban landscape

So you have your space set up along with your native bee hotels or hives — now to feed them! Most urban areas have a surprising amount of flowers and, in some cases, can be better than rural landscapes. At the same time there is an increasing focus on the 'green' infrastructure of our cities, driven by issues with climate change, stormwater, soil conservation and biodiversity.

But there is another key reason we are seeing more trees in urban areas across the country, and it is a guarantee that this trend will continue: urban trees increase property values. Not by a little bit, either. And not the trees on individual properties (although they matter, too) — it is the street trees on common land that increase values the most. Leafy, lush green streets increase the median house price anywhere from 2 to 4.5 per cent, and that equates to some serious money for homeowners.[1] There are lots of great reasons to have more trees in our cities, but the Australian obsession with house prices alone means that we are almost guaranteed to have urban forests throughout our cities and towns.

For urban beekeepers this is more than just a happy coincidence. It is a golden opportunity to influence the way these forests are designed to enhance biodiversity and create unique food experiences.

Flowering hedge of *Callistemon* at Parliament.

FLORAL RESOURCES: FEAST OR FAMINE FOR YOUR HIVES

It takes a lot of flower resources to feed a beehive — estimates range from 1 to 2 million flower visits per kilogram of honey. New beekeepers are often daunted by the scale of flowers required to produce honey, but it is worth breaking this down a bit. A full-sized tree or large shrub can have literally thousands of flowers, and the nectar within the flowers can recharge several times per day. Honeybees in particular have a system of 'scout bees' who continually search the landscape around the hive for new floral resources for the hive to gather nectar and pollen from. These are the two key resources that feed your hives, and it is essential that they can build up enough stores to survive.

Nectar is made up of liquid carbohydrates that are produced by plants as a reward for pollinators. Nectar forms the flavour base for honey, and different plants impart dramatically different flavours, colours and thickness to the honey produced from their blossoms.

Pollen consists of colourful granules of protein that contain the plant's genetic material. These are the 'packages' that bees and other pollinators deliver between flowers, but pollen also forms a critical food resource. It is the principal source of protein to the bees, and also imparts subtle flavour to the honey.

As I mentioned on p. 3, as a junior forester my class received a guest lecture on nectar and pollen resources by a commercial beekeeper who kept bees in native forests, and the main lesson he imparted was the critical role pollen plays in bee health. Many flowers produce nectar, but not all produce high-quality pollen, which requires both a high level of protein (above 25 per cent) combined with a diverse range of amino acids. Together, these form the building blocks that allow a hive to replenish their workforce. In particular, having access to a diverse range of pollen sources is critical for optimum nutrition. Much like our diet, bees require a range of amino acids, vitamins and minerals that occur in varying levels across different plant species. This is one area where cities can be far superior to country areas for bee nutrition — there is a wide range of plants grown as street trees, in parks and in backyard gardens.

Bees carry pollen in a variety of ways. *Megachile* species (Leafcutters and Resin bees) carry pollen on their abdomen, which the pollen turns bright yellow or orange. Honeybees carry it on their hind legs, making them look as if they have 'saddlebags' full of colourful flower pollen. Once inside the hive, the pollen ends up in different places depending on the species of bee. Solitary native bees mix it with nectar into a paste to lay their eggs into. Native stingless bees pack it into 'pots' full of pollen in the top of their hive. The most beautiful pollen storage is in honeycomb, where it is packed into cells near the brood, creating a mosaic of beautiful colour inside the hive. Pollen comes in myriad different colours, from drab green through to bright yellow and shades of orange, red, violet and even white. The result can be a stunning (and photogenic) example of pollen diversity in your area.

A lesser but essential plant resource for both honeybees and native bees is resin, which the bees transform into a multipurpose material called propolis. This sticky brown resin coats the inside of hives and nesting cavities, and is a powerful antibiotic. In a honeybee hive, it is used to seal the small gaps in the hive and also to imprison unwanted intruders. In a native stingless

City planter boxes, alive with my bees!

Pollen being packed into cells inside the hive.

beehive, resin is even more critical, forming the walls of the 'pots' that are used to store the honey. There is usually ample tree resin in urban areas for bees, so this is not something that beekeepers have to actively manage.

Assessing floral resources

Assessing the flow of nectar and pollen into the hive is one of the most critical skills that a beekeeper needs to develop. Bees must have enough resources to survive gaps in flowering times caused by winter, drought, fires and heatwaves. As a beekeeper, it is critical that you only take a surplus of honey, and even then only when the bees can readily replace it to reduce stress. Colony health must come first. The old saying is that you should look after the bees, and they will look after you.

The best tools for assessing floral resources are your own eyes and nose. The classic trick

Honeybee getting covered in her food as she forages.

of 'beelining' involves observing the bees leaving the hive to see which direction foraging workers are heading in. Bees will, over time, correct their navigation to head directly towards a source of flowers, and this will give you the direction. Observing the returning bees will also yield clues. Are they carrying pollen? What colour are their 'pollen pants'? Are they carrying mostly nectar (which is indicated by them landing heavily, and an intense honey smell coming from the hive) or pollen (colourful bags on their legs)?

While honeybees can forage out to 5 kilometres from their hive, and native stingless bees out to around 1 kilometre, most bees prefer to forage closer to home. Taking a mental note of where the bees are heading and walking around the suburb will often reveal a group of trees with a busily buzzing group of workers!

Where they have a choice, bee colonies tend to steer towards a single large resource, rather than foraging across lots of small shrubs. The key for urban beekeepers is to get to know your area so you can identify two key things:

Nectar flows: These are the peaks in floral production, when the nectar and pollen is rolling in. In Australia this can be spectacular, with high nectar-producing plants like Eucalyptus, Callistemon and Melaleuca putting out millions of flowers in the suburbs, parks and local bushland. Under ideal conditions, honeybees can fill a box in as little as a few weeks. You sometimes need to be quick to provide them with extra storage space!

Nectar dearth: These are the gaps in production when nothing is flowering, and the colony needs to survive on its stored resources. When in dearth, colonies will also be more defensive as they are protecting their precious, hard-won honey!

The peaks and gaps in nectar flows depend on a range of factors including climate, soils, recent weather, and the strength and skill of the colonies. That said, there is a rhythm to the year, and becoming attuned to the landscape is probably the single most challenging and rewarding part of becoming a beekeeper.

BEE-CENTRIC URBAN DESIGN

It is funny how a single comment can sometimes spark a new idea; it's the most fun part of being in a design team. When I was helping design the bee-friendly urban forest of the Ginninderry sustainable community, I made an off-the-cuff observation that most native

Beds of nectar-rich flowers can create 'islands' of food for native bees spread throughout the suburbs.

bees were truly local, as they lived in a 200- to 300-metre radius of their nest. At the time this was a flippant remark, but the urban design team turned it into a really important principle, using the smallest foraging range of the native bees in the area as a design unit.

It started with what seemed like a fairly simple question from them: how far does a bee fly? For the Canberra environment where I live, our best estimate was around 200 to 300 metres for the beautiful natives like Blue Banded bees (*Amegilla* spp.) and out to around 5 kilometres for the introduced honeybee. Being urban designers, their focus was on building communities based around the greatest need, and they naturally extended this to the bees that we were trying to support in the landscape. In this case, honeybees were the lowest priority — as a large, tough bee with a wide foraging range, they could handle wide spaces between clumps of resources. Not so the native bees; they needed smaller areas of habitat where food, water and nesting sites were all in a smaller radius.

Taking this rough foraging range for the native bees, the design team then went about creating connected 'cells' of habitat based on the native bee with the smallest foraging range. Not only did this approach ensure that each area of the suburbs had its own little 'hub' of native bees, the connected 'cells' created a network of habitat that allowed for the flow of genes through the wider population. In the process, a patchwork of green spaces was woven through the suburbs, creating something the community loved. For the developers, this approach was a financial goldmine, especially in the aftermath of the pandemic. The newfound appreciation for green space meant that these suburbs became the hottest property in the region — I had originally toyed with the idea of buying there, but quickly gave that up once I saw some of the house prices!

Urban trees: Not just for the birds and bees!

More and more cities are investing serious money into urban forestry, creating an extensive piece of green infrastructure that is constantly maintained. While a lot of the initial investment was done for beauty and some shade, the critical role that urban trees play in maintaining livable cities is becoming increasingly apparent.

Melbourne urban trees online

When the City of Melbourne created an online database of the city's street trees, they did something really innovative, with some unexpected social results. Managing street trees is a costly business and often generates a lot of complaints and reports of damage. Another common issue is community concern about replacement and renewal. To manage this, the local government created an online spatial database that allowed residents to log in to see the status of individual trees on their street, in effect giving each tree its own email address where reports of damage or disease could be sent.

The result was a whimsical exercise in unintended effects, as people started writing not just sterile reports to the trees, but heartfelt letters of concern, expressions of gratitude and even poems to the trees. Council staff showed a good sense of humour by replying in character, telling people how they felt from the perspective of the tree. It remains one of the most delightful ways to connect people to urban forests that I have ever seen, receiving national and even international coverage.

Urban flowerbeds provide beauty and food.

The key point I want to make is that we no longer see trees as just something pretty to put here and there — they are an essential piece of infrastructure for our cities, and we need to treat them as such. A big part of any infrastructure project is gaining as much use from it as possible, which for trees includes shade, forage for wildlife, honey production and a sense of community. We need to demand more from our urban forests, and recognize the myriad ways they add value to our lives.

Trees add more than just colour to a city: they also combat one of the most damaging impacts of climate change by reducing what is known as the 'urban heat island' effect. This is caused by hard, dark surfaces like roads and buildings absorbing heat from the sun and retaining it, creating a heat sink that can 'cook' the city during heatwaves. In this case, trees don't just help us. Studies of European landscapes have shown that forests seem to provide the best conditions for diverse pollinator communities, and seem to provide a buffer against the impacts of a warming climate.[2]

This isn't just a vague feeling, either. It is backed by hard measurement showing the impact that green leaves in the suburbs have in turning down the heat. Areas with a good coverage of leafy, shade-producing trees were found to reduce mean radiant temperatures by 4°C, and the heat of surfaces by up to 12°C.[3] Across a city, that is a lot of heat that is being kept out of our homes.

Trees and water-sensitive design help to cool down cities during heatwaves.

Urban forests with a theme

One of the things advanced garden designers do is to create 'rooms' within a garden, with a cohesive look and feel to them. Really well-designed urban forests are similar, with a look and feel to suburbs and locality that creates a sense of place.

Rooftop apiaries are beautiful at night.

I have tended to focus on the functional aspects of these urban forests, but we need to remember that they are primarily about aesthetics. Having a cohesive theme can be as simple as growing all the same species, such as tall, open-crowned trees in parkland, and then dense, leafy trees like Oaks along suburban streets. These can also create spectacular seasonal displays — the Jacaranda trees in Sydney and the red splash of colour from Flame Trees (*Brachychiton acerifolius*) are an iconic part of some suburbs, drawing in visitors from all over.

Having a theme also creates a sense of comfort and shared experience, a sense of belonging that comes from the familiar look and feel of local streets.

Creating a bee supermarket: The urban tree base

For feeding large numbers of bees, moths, birds and other urban wildlife there really is no substitute for a vibrant and diverse urban forest for producing year-round forage. The key issue for urban beekeeping is maintaining sufficient volume of production to create a viable surplus. Even small trees are an incredibly resilient landscape feature that can produce hundreds and even thousands of flowers, each a valuable source of nectar and pollen.

Trees are such an incredible resource due to their sheer power and resilience. A mature tree can draw enough water into its canopy to rival a fire hose, and the leaves act like thousands of solar panels. A chemical process converts water, light and carbon dioxide into the starches and sugars that sustain the tree, their pollinators (and eventually us). The oxygen we breathe, produced by trees, is fairly useful, too.

Bees are integral to how trees reproduce, and honeybees in particular evolved in forests and seem to do much better in these environments. Even trees that don't directly feed bees through nectar are useful, with a lot of wind-pollinated trees providing resin for propolis and nesting habitat.

The joy of being in the middle of an orchard in full bloom, with the hum of your bees filling the air.

Managing an urban forest is an incredibly complex undertaking in practice and the potential liability of having trees close to people, cars and houses means trees must be constantly checked and maintained. Trees have essentially two lifespans in an urban forestry context. First, there is the biological limit of life for a tree, which can be as little as a decade for some large shrubs and exceed several centuries for others. However, for urban forests the safe working life of a tree is the critical consideration — this is the lifespan before the tree starts to become unsafe, dropping branches or risking sudden failure. As a result, trees are often planted and then replaced in groups in a continual cycle. This cycle of replenishment presents a big opportunity for beekeepers and native bee advocates. In the past, tree species were often chosen based on their aesthetic appeal and ease of maintenance. These are still important, but there are lots of beautiful, low-maintenance species that are also a wonderful resource for pollinators.

In developing the updated planting guide for Canberra, I and the other contributors considered the following pollinator-friendly attributes to supplement the traditional considerations of form, cost and maintenance:

Native species

While I am far from a 'natives or nothing' advocate, plant species that are native to a region should always be the first to be considered. The local species will be adapted to forage from them, and native plants are intrinsically adapted to the local microclimates. However, with the increasing progression of climate change, urban forest managers are starting to look at the future environments that trees will need to survive and thrive in. Unfortunately, a certain degree of climate change is now 'locked in' due to the inertia in the climate, so we are likely to see hotter and drier conditions. Equally, just because something is native to the

Tough natives like this Bursaria provide resources for a wide range of bees, beetles and butterflies.

region doesn't mean it works in an urban environment — riparian species, for example, may simply not have enough water.

Flowering period

Generally, longer flowering periods are better as this provides resources over a longer period, which allows for native bee species to provision multiple nest chambers. It also provides a longer nectar flow to fill honeybee hives. Equally important is to pick known gaps in production and use strategic plantings to fill these.

Productivity

The volume of nectar and pollen from species varies widely. In our region we have some species that only flower intermittently, and even then, only produce small amounts of nectar and pollen. On the other end of the scale are the famous honey-producing trees such as Yellow Box (*Eucalyptus melliodora*), with each tree producing thousands of nectar-laden flowers that can fill a beehive super (box for surplus honey) in less than a week. Equally important is the production of high-quality pollen, which is a critical resource for native bees and honeybees alike, providing the protein to build baby bees.

Size

This is a surprisingly important consideration. Unfortunately, some of the best honey-producing trees are also the largest, which tends to limit where they can be placed. The star performers in urban environments are increasingly the smaller trees and large shrubs such as *Callistemon*, which are small in stature but can flower twice in a year with nectar-laden, bottlebrush-shaped flowers that give the genus its common name.

Eucalypts are beautiful and useful street trees.

Bottlebrush flowers are laden with nectar, loved by the birds and the bees!

Replacing lost natural resources

This is one of the big advantages of cities, which often have access to water, relatively good soils and protection from major fires and floods (sometimes). When we were providing input on the Canberra street tree plan there had been a mass dieback of Manna gum (*Eucalyptus viminalis*) in the wider landscape. The exact reasons for this are still a mystery, but winter nectar and pollen resources were lost over thousands of hectares in the surrounding landscape. Looking at these wider landscape problems can help us create refuges for wildlife.

STOCKING UP THE HOME PANTRY: BEE-FRIENDLY GARDENS

The trend towards bee-friendly gardens has been like the trend towards beautiful, artistic restaurant dishes — adding colour, beauty and pleasure to our lives. There really are no downsides here, as creating a community garden or urban parkland with year-round flowers is a lovely way to brighten up our cities, while at the same time providing critical supplementary forage. However, not all desserts are created equal, and flowers are no exception.

As my waistline will attest, a lot of small desserts still adds up to a lot of food. Bad for me (as someone keeps sneaking into my house and making my clothes smaller) but fantastic for

A dense and long-flowering urban garden bed.

A diverse bed in a community garden with a variety of long-lasting flowers.

the health of bee colonies. Canberra is a city of gardeners and there has been a very noticeable lift in the number of people growing flowers specifically for bees, butterflies, birds and other pollinators.

There is one group of bees that has been particularly favoured by urban parks and gardens, and these are the native bees. These colourful, clever creatures typically have a home range that is a fraction of that of the honeybee, so the bees visiting a vibrant, native-friendly garden really are your local bees. More importantly they are incredibly useful, with several of them being critical pollinators for flowers that are generally poorly pollinated by honeybees.

Discovering native bees

As a young forester, one of the first native bees I became fascinated by was the Blue Banded bee, and the one that lived in my local forests and woodlands (*Amegilla cingulata*) in particular. This is one of the most common bees in Australian gardens, and other members of *Amegilla* are found in many parts of the world; I was particularly pleased to come across a Blue Banded bee while drinking beer in a bar above a river. Normally this would be a typical experience — except that I was in Uganda overlooking the White Nile. This beautiful bee was happily foraging in the flowers next to the bar, and was virtually indistinguishable from our Australian species.

Blue Banded bees are critical pollinators for a range of native plants, but they aren't particularly fussy and love Lavender and other introduced plants. For gardeners they are particularly important — they are a critical buzz pollinator for Tomato plants. While some plants will give a gift of pollen to any old insect passing by, some plants have developed an ingenious system to make sure that only their specific pollinator will pick up the goods. The flowers hang downwards and will only release pollen when vibrated at a specific frequency — sort of like having to sign to pick up a parcel. If you look closely you will see the Blue Banded bee grab hold of the flower and run its wings in neutral, sounding like a miniature taser going off. There will be an immediate shower of pollen from the flower as it is shaken loose. The downside is that, as native bees decline, generic pollinators like the honeybee are just not up to the job.

There really is no substitute for a diversity of native pollinators to maintain our gardens; while they may not make as much (or any) honey, in terms of beauty and the productivity of our home vegie patches they are the superstars of the productive garden.

Blue Banded bees glean nectar in a variety of ways, even stealing from the sides of tube flowers. (Credit: Dr Kit Prendergast)

The best thing about making community gardens and parklands more bee friendly is that even small things really add up, and it really is an example of an action where every little bit helps. It is also completely compatible with beautiful, interesting green spaces for us humans, with an emphasis on year-round flowers, hidden habitat areas and water features.

Picking gaps in production

As discussed on p. 73, bees forage in a radius of 200 metres to 5 kilometres from 'home', depending on the species. So in most cases it is impossible to sustain colonies with just the flowers growing in your community garden; however, this is the area you have the most control over. The best thing you can do for your local bees is to figure out where the critical gaps in flowering times are, and shift the species mix in your garden to fill these gaps. In my local area, the main gap is in autumn, which is when both natives and my honeybees are doing their last bit of food collection before winter sets in. So, for me, maintaining a set of plants that produces abundant nectar and pollen during the autumn months makes sense.

Figuring out when the floral gaps are takes a bit of detective work in the local landscape. With honeybees it can be apparent during inspections if no fresh nectar is evident for a particular period; this can be a clue for what to plant for the next season. For native bees, it is more about observation — are there flowers in bloom? If so, are they actually being visited? Some non-native flowers are loved by our native bees, but others are not so palatable. This is where taking part in the citizen science programs mentioned later (see p. 218) can be really useful.

BEE-FRIENDLY PLANTS

The list of bee-friendly plants could fill a book on its own, and literally does in a few cases! These are fantastic resources and are among my favourite books. However, to get you started here are some classic plants that every bee enthusiast should make some space for.

Borage (*Borago officinalis*)

Top of the list as far as the bees are concerned, this species grows really easily with minimal care, self-seeding year in year out, with beautiful blue flowers all the way from spring through to autumn. Borage flowers are a magnet for bees, and have one of the fastest recharge rates, replenishing their nectar every few minutes. They're dead easy to grow — just scatter seeds

Borage is the classic bee-friendly flower.

onto some bare ground and rake into the soil, or establish in peat pots that can be buried with the seedling. You can even use the beautiful blue flowers as cake decorations, and they have a refreshing cucumber flavour. One of my favourite tricks is to freeze the flowers inside ice cubes to add a splash of colour to summer drinks.

Sage (*Salvia* sp.)

If I was stuck with just one genus of plants to fill a pollinator-friendly garden with, these would be my pick. Salvias are easy care, drought tolerant, long-flowering and absolutely loved by native bees, honeybees and butterflies. The highest number of Blue Banded bees I have ever seen on a single plant was clustered around a Bog Sage (*Salvia uliginosa*). There is no need to be too picky with this whole genus of plants — just head to a nursery, pick the colour you like most from the hundreds of varieties and get planting. They are all a hit with pollinators and are a good example of a non-native plant that can really help support native bees.

Catmint (*Nepeta cataria*)

This is another beautiful groundcover that is loved by both honeybees and native bees. Beautiful grey-green foliage is complemented by masses of blue flowers throughout summer, making it a favourite for gardeners as a border plant. I grow it throughout my garden for its other properties — it is a potent cat drug, making your furry companion high as a kite and extra playful. While honeybees will regularly visit, this plant is a firm favourite of Blue Banded bees.

Loved by bees and my cats — Catmint is another classic.

Raspberries (*Rubus idaeus*)

This is really only for people in cooler climates with the room to grow them, but Raspberries go together with bees better than almost any other backyard crop. For starters, Raspberries don't travel that well and are fairly expensive to buy, so you get a lot of bang for your buck when growing them. For bees they are useful in a couple of ways. The most obvious is the flowers themselves, which put out a large amount of nectar and are highly sought after; this is good because the canes require multiple bee visits for decent fruit-set.

Once you get your Raspberry patch established you will need to keep cutting out the old canes as they die back. Save these, and see p. 14 for how to use them to attract even more bees to the garden.

Lemon Balm (*Melissa officinalis*)

Lemon Balm is dead easy to grow, looks great and comes out in masses of white flowers through summer that bees of all varieties absolutely love. It is a self-seeder, so it is basically zero maintenance other than trimming back every now and then. This plant is as useful in the kitchen as it is in the garden, with a lemon fragrance that can be infused into tea, sweet treats and even salads. The Greek name for this plant translates to 'bee balm', so this has been a star performer for centuries.

Red Bottlebrush (*Callistemon viminalis*)

This is a beautiful tall shrub to small tree that is fast growing, easy to care for and has attractive dark-green foliage. They are even starting to be used as a street tree, as they don't shade out solar panels. The common name comes from the shape of the flowers, which are like an old-fashioned bottlebrush, and have dozens of individual flowers bursting with nectar from October through to February.

For bees, birds and other pollinators they are a fantastic food source. In the original Aurecon office apiary, the bees would swarm the large shrubs that were grown next to the office, and now there are literally hundreds of plants in the Parliament gardens, including the hedge fringing the main entrance. It seems somehow fitting that a significant portion of the honey in the hives comes from the front entrance of the building.[4]

Raspberries need lots of bee visits, but the flavour of homegrown berries is worth it.

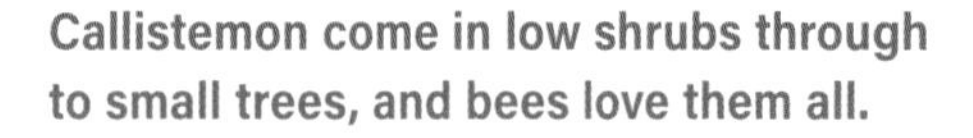

Callistemon come in low shrubs through to small trees, and bees love them all.

The versatile, beautiful and hardy *Hardenbergia*.

Happy Wanderer (*Hardenbergia violacea*)

One of the best 'gap fillers' that is beloved of native bees is this hardy, low maintenance native species. The dark-green foliage gives the perfect backdrop to the striking purple pea flowers that come out from autumn through to spring. Equally happy climbing up a fence, scrambling over a rockery or spreading as a groundcover, Happy Wanderer provides critical flowers during late winter and early spring. In addition to covering the soil as a 'living mulch' this species is leguminous, meaning that it fixes nitrogen into the soil, providing nutrients for other plants. The leaves can even be boiled into a sweet tea. A useful native that I always have in my garden.[5]

Alyssum (*Lobularia maritima*)

This plant is the ultimate gap-filler in every sense of the word, able to squeeze into gaps in garden beds and flowering all through winter and into spring. I have observed a wide variety of pollinators happily foraging on these plants, particularly the smaller native bees that are rarely accommodated in gardens. Dead simple to source and establish, this is the base for many of the pollinator-friendly seed mixes you can buy, and it takes minimal care to establish and maintain.

Gorse Bitter Pea (*Daviesia ulicifolia*)

This is a compact native shrub that produces masses of yellow and orange flowers, which appear on the bush from late winter and then persist through spring. While not especially

Alyssum is a great garden gap filler.

The spectacular Grass Tree flower spike.

high in honey production, it produces abundant pollen and nectar right when native bees are emerging from their hibernation. Perfect for understorey planting of your front garden, as the slightly prickly leaves deter intruders.[6]

Grass Tree (*Xanthorrhoea glauca*)

If you are looking for a spectacular garden specimen this is the one I would recommend. However, a word of warning: unless you are very rich this is going to be a long wait! Grass Trees are super-slow-growing natives that build up their resources over years and even decades. You can buy established plants for several hundred to several thousand dollars, but smaller, affordable plants will take several years before they are ready to flower.

So if they are a hassle to grow, why bother? The answer is the beautiful, weird grass leaves capped off by the frenzy of pollinators visiting the flower spike of a Grass Tree! Rather than putting out a single flower, or even a bunch, when Grass Trees flower they set out a spike with thousands of nectar-laden blossoms, and they have so much nectar that you can drink it right off the plant. If you ever get a chance to do this, don't hesitate, it is an amazing taste sensation!

One flowering is finished, don't throw the flower spike away — this is a critical nesting resource for native bees including the stunning Emerald Carpenter bee. This was one of the great utility plants for Indigenous Australians, with the silica-edged leaves sharp enough to work as knifes to slice meat, the resin from the base of the trees used as superglue, and the spent flower spikes used as fire starters. They really are fantastic, and more people should grow them.[7]

BEEKEEPER-FRIENDLY PLANTS!

It is all very well for us to use our garden to look after bees, but what about the beekeeper? While native bees are remarkably low maintenance and almost never sting, honeybees are not always so forgiving! A few strategic plantings of useful species can really help with the function of your garden. I always have a few of these utility plants handy, as they are extremely useful during inspections.

Ribwort Plantain (*Plantago lanceolata*)

This is basically a weed on roadsides and disturbed ground, and it is probably growing somewhere out of the way in your yard or nature strip right now. Often overlooked, it has one fantastic property for beekeepers: it is a great sting antidote. It is very simple to use: just chew up some of the leaf and press the pulp against where you were stung for almost instant relief. You don't need to do any special cultivation for this herb — just know where it is growing near your hive and don't pull it out! I used to discount the flowers of this plant as they are scraggly and don't really look like flowers at all, just a sort of spike with some petals sticking out. But then I noticed that the bees were always on this plant. Some flowering plants bees will occasionally visit but mostly wander past, but this one seems to have a magnetic quality as far as bees are concerned. I am still dubious about its value in terms of volume of nectar, but the bees seem to love it so that is good enough for me!

It is not a weed if it feeds your bees and heals stings!

Handy in the apiary and the kitchen.

Winter Savory (*Satureja montana*)

This is another useful bee-sting antidote, and the advantage of this herb is that it is both hardy and very handy in the kitchen. In my experience, Winter Savory is a sting antidote stronger than Ribwort Plantain, and providing faster relief than Ribwort, although the latter seems to provide more lasting relief. In the kitchen, Winter Savory provides a bit of peppery bite to bean dishes, and is also good for seasoning meats — it goes really well as part of breadcrumb stuffing mixes. As a nice bonus, it is relatively easy to grow and comes out in masses of white flowers that bees of all varieties seem to totally love. Very much the all-purpose utility plant for beekeepers, this is one that I grow next to where I put my suit on as I get ready, so that I can quickly grab a sprig if I get tagged.

Safety note

These sting antidotes are mild natural remedies, suitable for mild irritation only! For actual medical problems, you need actual medicine, not mild folk cures. If you have any medical concerns, get medical advice pronto!

SNACK FOOD: SMALL-SPACE CITY GARDENING

Just like the tasty snackables crammed into a lunchbox, cities often have small spaces that can be filled with flowers for a quick pollinator's pick-me-up on the go. Small spaces have recently become more popular as beautiful photographs of roadsides, roundabouts and even bus shelters blooming with wildflowers attest. While these are often promoted as a 'save the bees' initiative, the real reason is money — it costs a fraction to establish wildflowers and let them bloom compared to intensively mowing these areas. As a nice bonus, it actually does work as advertised, with recent research proving the worth of these roadside plantings, especially when they contain remnant vegetation and connections to other habitat patches.[8]

The key requirement for planting these micro-gardens is that the plants are very much set-and-forget, while still helping out whoever happens to pass by needing a pick-up of nectar and pollen. Natives like *Hardenbergia* and self-seeding exotics like Catmint work really well in these spaces, as they support a wide range of species and require almost no care.

Planter boxes can hold a surprising amount of plants with a bit of layered planting.

Some of the best examples of these small spaces that I have seen in cities are planter boxes near cafés. Too small to be a fully manicured garden, these raised boxes are perfect to rest your coffee cup on while checking the news, and are often still big enough to hold a decent number of plants. A few signs to let people know about bees, and a bit of urban art, and you have the perfect island of green forage in among the concrete.

BEE HIGHWAYS: JUST AS IMPORTANT AS OUR ROADS

So we have all the plant components outlined above, but the really critical thing is that these are not randomly spread throughout the city in a disjointed way. Increasingly, the best practice approach is to create 'bee highways' that provide a continual corridor of forage and nesting sites. This is where governments need to work with communities to create a planned landscape that creates diverse pollinator communities, which includes bees, moths, flies and even some native wasps (the nice ones).

This is particularly critical for native bees, as their home range is so much smaller. While honeybees will forage and travel in excess of 5 kilometres from their hives, native bees are real home-bodies, preferring to stick to an area of only a few hundred metres from their nest. As a result, big gaps in foraging and nesting material

This community garden mixes fruit trees in with native trees.

will isolate populations, creating the potential for inbreeding.

Linking these disparate 'islands' of vegetation won't just help the bees — these are critical movement corridors for birds, bats and other insects, all of which are important parts of the environment. At this larger scale it is viable to collect local provenances of native plant seed, conserving local flora at the same time. Two of the ways that this has been approached at city level include planning guides that specifically reference flowering times, and pollinator corridor declarations.

Planning guides

Many cities have a systematic approach to planning their urban forests, which includes schemes for a diversity of tree and shrub species within broad themes to suit planting situations. In Canberra this guide was revised in 2019 to include specific consideration of pollinator habitat, including the flowering period of plants, whether they provide nectar, pollen or both, and what sort of fauna utilize them. The important element at a city scale is that this needs to specifically encourage a continual supply of nectar and pollen in the environment within each suburb or area. Just planting lots of spring-flowering plants might be nice, and will be great for the species that are active at this time, but it will be useless if the urban forest is a green desert for the rest of the year.

Pollinator corridor declarations

These can be formal policies or community-based initiatives, and are usually based on mapping existing pollinator habitat, protecting these critical 'islands' in local planning schemes and then identifying the gaps that need to be filled with new habitat.[9] Creating these corridors generally needs to involve a combination of plantings of forage, nesting habitat and water throughout public land and private gardens. Some cities have even created pollinator-friendly garden schemes that allow landowners to publicly display their commitment on their letterbox.

One of the really nice things about this approach is that the native bees in the gardens will provide a more diverse pollinator community, increasing yields of backyard vegie patches, fruit trees and flowers. They are also a powerful way to engage communities in seeing the landscape as a whole system rather than just their little patch. At the same time, the diverse community of bees has become something of a competition in some areas, and it would be fun to see a bee version of the Tidy Town competition spring up.

Love the blend of artwork and garden in this city planter box.

GUERRILLA GARDENING

With our cities often a drab, concrete jungle, it is easy to just accept that this is the way it has always been, and always will be. However, there is a subtle and fun way to subvert this, by taking drab, neglected ground and creating something beautiful and productive. The term 'guerrilla gardening' was first coined in the 1970s when the Green Guerrilla group led by Liz Christy started a community garden on a vacant lot in New York City. The garden remains to this day, and spawned a global movement that has reached cities around the world.

However, the idea is much older, going back to the mid-1600s when Gerrard Winstanley and the Diggers began taking over vacant land and growing crops to share among the community. This did not go over very well with the landed gentry, who liked their serfs beaten down, so they hired armed thugs to … beat them down. Luckily, armed thugs are harder to come by these days (and the serfs are more uppity), and both governments and the community have taken to the idea of filling our cities with flowers and vegetables.

The key to doing guerrilla gardening successfully is to find patches of neglected land that nobody cares about, and then to work in a subtle and clever way so that, by the time anyone notices, there is already something interesting and fun in that space.

It is really important to not actually break the law — trespass or vandalism is never okay and is ultimately counterproductive. You want to inspire and delight in a fun way, not cause outrage. A critical part of this is to not create new problems, like weeds or trees on top of water pipes — we want colour and food for pollinators, not repair bills.

For me, guerrilla gardening is all about finding the forgotten patches of common ground that nobody has thought to use, and turning those patches into something beautiful and useful. In the case of many Australian cities, the nature strip out the front of your house is public land that you are meant to maintain, and is often mown exotic grass. This is just the done thing, but why? Wouldn't it be more fun to cover it with wildflowers?

This is all about small space gardening, turning forgotten patches of ground into a colourful and useful part of the cityscape. Here are some ideas.

Seed bombing

This is the ultimate in 'set and forget' planting, involving seed balls of equal parts artist's modelling clay and compost infused with seeds. The dried ball protects the seeds from heat

This courtyard between office buildings has plants that flower at different times, providing forage over an extended period of time.

and wind and ensures that they stay dormant until enough rain falls to dissolve the ball, creating a small pile of soil that the seeds sprout from. Perfect for having a few in a bag or tin to throw into abandoned ground, as they have all they need to create a small patch of flowers. These are just for abandoned ground in urban areas — never for remnant bushland patches. Leave seeding these areas for your local Landcare group (or join in yourself)! See p. 95 for how to make your own seed balls.

Wildflower verges

These are becoming really popular due to the beauty (and cost savings) of flowers instead of mown grass. On a large scale, wildflower verges are largely the realm of local government, but in most places the management of the nature strip out the front of your house is your responsibility … and your opportunity. Even if you can't bear to completely give up a manicured lawn, it doesn't have to be grass: both thyme and chamomile come in low-growing lawn varieties that create a beautiful carpet of flowers.

Small patch gardens

There are always little corners of ground in the suburbs where there is enough soil for some wildflowers to take hold, creating some colour among the concrete. Choosing the right spot can be tough with our harsh summers, but look for areas that get morning sun and some shade in the afternoon. Establishing these patches can be as simple as pulling out any weeds, raking over the soil and planting some herbs that will run to flower.

Community plots

Almost every city has community gardens now, nestled in among the suburbs on ground that nobody else wants. I used to run the Railyard Garden in Queanbeyan, which was established on an abandoned railway maintenance yard. This provided low-cost ground for people from the surrounding area to grow vegetables, flowers and herbs for the kitchen. There can sometimes be a waiting list for popular community gardens, but it is a pretty fantastic way to connect with your local community.

Community gardens commonly emphasize organic and companion-planting methods, which are a very bee-friendly way to garden.

Seed ball recipe

Here's what you need to make your seed balls if you want to do the seed bombing covered earlier on p. 94. If you use one cup to measure below, these ingredients should make around 20 to 30 seed balls, depending on how small you roll them.

5 parts artist's clay
3 parts sieved compost
1-2 teaspoons of slow-release fertilizer
2-3 teaspoons pollinator seed mix suitable for your region (remember: don't include anything that could become a weed in the environment)
Water to moisten

Mix the dry ingredients together and then moisten with a hand sprayer until the clay can be formed into balls about the size of marbles. Roll between your hands and then lay out on sheets of newspaper to dry in a shaded area. Then just throw the balls at some waste ground in your neighborhood and wait to see what emerges!

Components of a seed ball.

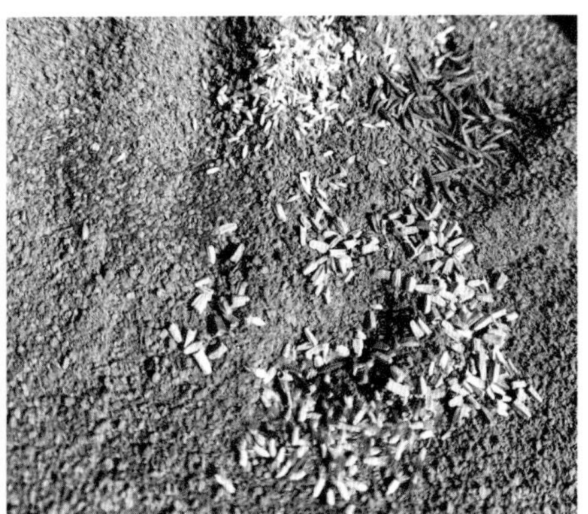

Adding (non-weedy) seeds to the mix.

Finished seed balls, ready for drying.

Native wildflowers like this *Hardenbergia* should always be on your seed list.

There are a few basic principles that apply to all of the above approaches.

Native is best, and never weeds

Wherever possible, try to use native wildflower seeds that have been sustainably collected. This can be a great way to support local seed collection companies, and you can be sure that it is a locally adapted variety that will support native bees. If these are not available, then pollinator seed mixes can be a really good option, especially for inner-city situations away from nature reserves. Never use declared weeds, as these can be invasive and cause ongoing problems for native wildlife.

That said, native bees are definitely partial to some exotic species, with a particular love for small-flowered forms of Catmint, Sage and Rosemary, as well as the Dandelions in your lawn!

Mix it up

Don't rely on just one type of flower; go for lots of diversity. Pollinators need a variety of flower species for optimum nutrition. Try to not get 'tunnel vision' about what you want to achieve in terms of colour or form — try a mixture of herbs that will run to flower like Basil, traditional wildflowers and even edible flowers. There is a style of French cottage garden called a potager garden that mixes fruits, flowers and vegetables together to create a form of edible art.

Artistic expression

This is about expression and beauty, so don't miss an opportunity to incorporate a bit of art into the science of feeding pollinators! Patterned plantings, pop-up sculptures and even temporary murals could be on the cards. Even a little bit of art will make people stop and take notice of what is happening, and you never know who you might inspire. A simple native bee hotel is an inexpensive point of interest and lends itself to a colourful paint job to bring a disused area to life (literally).

FOOD FORESTS

The bigger and more organized sibling to guerrilla gardening is the food forest. This is a planned urban orchard that is owned by the community and creates free food for people to try. This is a chance to give the community a hands-on experience of growing their own food. There are few culinary experiences more magical than fruit fresh off the tree, and these forests are an opportunity to grow unusual varieties that folks may not have experienced before.

There is a lot that goes into growing healthy fruit trees, including intensive site preparation, protecting the young trees, pruning and managing pests. This is all before you taste a single piece of fruit. It is totally worth it, but works a lot better when there is a community to help manage and maintain the orchard. The benefit is really worth it, though — not only will the trees create beautiful shade, but they also provide a sheltered microclimate that allows all sorts of fruit, flowers and vegetables to be grown underneath the trees. At the same time these have the potential to create a sense of community that goes beyond the produce.

Food forests like this one mix flowers, native trees and fruit into a beautiful and useful community resource.

A blend of wildflowers, food and commuinity green space.

6

Hive types

Try something a little different

The type of hive you use is an intensely personal choice, and there is no 'one size fits all' hive design, with each having its advantages and disadvantages. In our training apiary at Canberra Region Beekeepers (where I run courses) we deliberately keep a wide variety of hives, and I do the same at the Parliament apiary. This lets people get some hands-on experience with different hives, and they often come away with a different view of what they want to go with. I have been fortunate to work with every configuration of hive available in Australia, and I have definitely developed some favourites! One of the big advantages of being a small-scale beekeeper is the opportunity to use hive designs that wouldn't necessarily work in a commercial situation but are ideally suited for backyards.

In terms of legal obligations for beehive design in Australia, there are only two absolute requirements:

- **The hive must have removable frames so that the brood (baby bees) can be inspected for disease.**
- **The owner's registration number must be printed on the outside of the hive.**

Parliament apiary, with multiple hive styles. (Credit: 5 Foot Photography)

Everything beyond this is personal preference, and I have summarized the pluses and minuses of each below.

There are two basic sizes of hive used in all the systems discussed in this chapter:

- **Beehive: Describes the full-sized hive capable of containing a colony in the long term, meeting all their needs in terms of space and protection from the elements.**
- **Nucleus hive: This is a miniature version of the hive that can usually take four to six frames of honeycomb. Nucleus hives are used to catch swarms, transport queen bees and split a hive to create a new colony.**

All beekeepers who work with honeybees need the full-size hive, but so often I see beekeepers who don't purchase a nucleus at the same time. Every beekeeper should have a spare nucleus hive (or 'nuc') on hand at all times; they really are an incredibly useful piece of gear. They are the secret to easier beekeeping with honeybees — get one, or even more.

So what sort of hives are out there? Let's look at some of the less standard hives that you might like to try out.

THE BULLETPROOF HIPPIE: KEEPING BEES IN THE WARRE HIVE

- → Creates resilient bee colonies on natural comb that can resist extremes of heat and cold.
- → Smaller boxes make for less heavy lifting.
- → Unique look to the hive, often beautifully painted.
- → Commonly handmade, would suit a hobbyist carpenter.
- → Slightly less honey, but more wax.

The first hive that I ever built myself was a Warre hive, based on plans off the internet. It was pretty terrible — rough corners, finger joins that were too big (lots of filing) or had gaps, and was not entirely square. While I had to hide my face in shame every time my woodworking friends saw these early creations, the bees didn't care and happily moved in. With experience and a bunch of better tools I started turning out Warre hives good enough to sell, but I kept the old boxes and still use them as swarm traps today.

Warre hives are my go-to system for really tough sites like urban rooftops, where the bees have to contend with a combination of heat, cold and high winds. These are a halfway house between the control of a Langstroth and the free-form of a top-bar hive. As a sort of jack-of-all trades, Warre hives are versatile, but there are some tricks you need to use to get the most out of them.

The Warre hive is an attractive, resilient hive design.

Originally created as 'the people's hive' by French monk Abbé Émile Warré, this hive mimics the size and honeycomb sequence that Warré observed in natural tree hollows. The one compromise on natural beekeeping is that the smaller boxes have frames inside them to allow for brood inspections and manipulations. The frames are left open at the bottom, providing (in my opinion) the best

The stunning Indigenous painted hive at the Aurecon office. (Credit: Kris Arnold)

of both worlds — you get the natural comb structure of a top-bar hive, but the capacity to manipulate colonies like a Langstroth hive. They are also extremely convenient for catching swarms!

The Warre hive consists of three to four (sometimes more) boxes stacked on top of a base with an entrance hole, with each box containing eight open frames for the bees to build honeycomb onto. What sets the Warre system apart is its quilt box and roof system. A hessian cloth is placed on top of the colony and on top of this goes a quilt box filled with wood shavings, straw or other fine material. This acts just like the quilt on your bed, keeping the colony warm in winter, and absorbing excess condensation. On top of the quilt box goes a roof, which seals the entire top section; the only ventilation is from the entrance hole below. The roof has a gap underneath it, shading the hive and allowing for natural ventilation during hot weather.

The other key difference is how boxes are added — always on the bottom of the hive, called nadiring or under-supering. This mimics natural comb progression that would occur in a tree hollow, with the bees initially building brood cells for baby bees, then extending the brood nest down as more honeycomb is built. As the baby bees hatch, the cells are backfilled with pollen and then finally honey. By the time the hive is stacked four boxes high, the top two are usually filled with honey and ready for harvest. This provides the two key advantages for natural colony growth: continual production of fresh comb, and bees exclusively building onto wax comb.

To harvest honey you simply take off the top box, shake the bees off the comb, cut out the

Classic comb structure of a Warre hive.

Oops, they don't always build straight with this hive design — these girls preferred to go diagonally!

honeycomb into a bucket, leaving a small strip of wax at the top, and then place this box of now empty frames underneath all the others. This re-starts the sequence as the bees keep building fresh comb downwards.

A Warre hive is a really elegant system that closely mimics what happens in nature, while still allowing you, as a beekeeper, to check for disease and harvest honey. Hygiene is maintained by bees always building onto fresh comb, and the queen is allowed to freely roam throughout the hive.

The Warre hive excels in terms of catching and nurturing swarms, and is one of the easiest hives to establish. I have a lightweight top and base that goes onto a single Warre hive box that I use to catch swarms in, and this then forms the nucleus for the new hive. The biggest advantage of this system is that the swarms establish in the box that they will live in — no double-handling to move frames into a new box. The swarm also has ample space to build natural comb, which they have a strong instinct to do. Swarms have a lot to do getting established, and the relatively compact space inside the Warre hive, plus the quilt box on top, helps them keep warm. For particularly large, vigorous swarms, two boxes can be stacked on top of each other, essentially forming an instant beehive!

As with the slightly crusty exterior of your average hippie, the Warre is an inherently messy hive — the fact that the bees are allowed to 'freestyle' means they will build comb between boxes, and the more open nature of the frames means that cross-comb, where the bees build bridges between combs, will occur. If left to their own devices, it is not uncommon for almost the entire box to become a tangled mess of honeycomb. It is more important with this hive than almost any other to undertake regular inspections to trim excess comb and bend 'crazy comb' back into the straight frames, especially during a strong nectar flow.

LETTING THEM BEE: TOP-BAR HIVE

→ Beautiful looking hives that are easy to inspect and generally gentler on the bees.
→ Eliminates heavy lifting entirely, and a convenient working height.
→ Difficult to migrate — suited for home gardens.
→ Limited standardization — generally cannot swap parts between hives.
→ Slightly less honey, but more wax.

Top-bar hives get their name from the wooden bars that the bees build their comb off. Instead of confining frames, the bees are allowed to build natural comb downwards into a long cabinet, which is commonly raised on legs to a comfortable waist-height. The colony builds their combs sideways, and inspections are carried out by separating a single 'top bar' at a time, lifting it out to inspect the comb.

This is a very gentle way to keep bees, with inspections only opening a small part of the hive at any one time. This often rewards you with easier inspections and less aggro from the colony, as you are mostly working with foragers and hive bees, well away from the guards at the entrance. The design also promotes gentle, gradual movements, which helps keep the bees happier during inspections.

It isn't just gentle on bees: it is quite gentle on the beekeeper, too. There is basically no heavy lifting with this style of hive — the most that a single bar can put on is 2 to 3 kilograms of honey, and you only need to move them one at a time. Harvesting honey is also well suited to a community garden for occasional harvesting. Once the honeycomb is fully capped, you just brush off the bees and cut it away from the bar, leaving a starter strip of remainder comb of a few centimetres. You don't need to harvest an entire box; any leftover honey can simply be left in the hive as reserve food for the bees.

Top-bar hives represent the limit of allowing bees to manage their own space while still meeting the legal requirement to have removable frames. The hives are often beautifully handcrafted, and in my experience, are easier and less intrusive to inspect than other beehive types. Instead of rigid frames, the colony is allowed to build natural comb off a set of horizontal bars, with new bars added sideways against 'follower' boards that keep the ends of the colony confined and warm.

The main disadvantage to this style of hive is a generally lower capacity to manipulate the colony. Due to the handcrafted nature of these hives you are unlikely to be able to obtain a spare hive from a beekeeping supply store. Likewise, nucleus hives and spare top bars need to be custom made. In reality, this is not a problem for most beekeepers; with some basic woodworking skills you can make replacement parts cheaply and easily.

One variant of the top-bar hive that works really well is a horizontal Langstroth hive. Essentially, this arranges the three boxes of a standard Langstroth hive horizontally, so that you don't have to

Top-bar hives are a nice, gentle way to keep bees.

Lifting one comb at a time means no heavy lifting with this design.

Natural comb filling with honey.

Combs are built sideways throughout the hive.

do any heavy lifting, but with the advantage of standard frames. These need to be carefully constructed, as the weight of the full hive can exceed 100 kilograms, but they have the added advantage of using Langstroth frames, meaning that you can use standard gear from commercial hives. This solves one of the biggest problems with top-bar hives, in that their handcrafted nature means that getting spare parts can be a hassle.

THE RESILIENT CLIMATE HIVE: SLOVENIAN AZ HIVE

- Beautiful, intricate design that is intended to be built into the wall of a shed or house.
- Often features painted entrances with cultural and biblical scenes.
- No heavy lifting – cabinets where frames are kept, like books on a shelf.
- Best protection against extremes of heat and cold.

They say that old tricks are the best tricks, and the Slovenians were the original professional beekeepers. While everyone else was still treating beekeeping as an art to be passed on through folklore, Slovenia developed the first systematic way to educate new beekeepers. The genesis of this was with Anton Jansa, who created the first syllabus for training beekeepers as a profession, and whose birthday is now commemorated across the globe as World Bee Day.

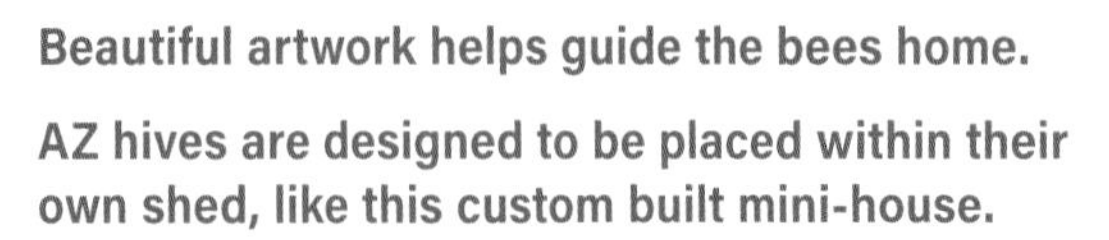

Beautiful artwork helps guide the bees home.

AZ hives are designed to be placed within their own shed, like this custom built mini-house.

The AZ beehive represents centuries of refinement in design and is a beautiful, intricate system that was developed to allow bee colonies to survive and thrive in the harsh, mountainous conditions of Slovenia. The hive is arranged like a set of drawers that opens from the back. The frames of honeycomb sit on metal rails and are pulled out like books on a shelf during inspections. The beauty of this system is that only one frame is removed at a time, and there is no heavy lifting to pull boxes apart. This hive shares a big advantage with the top-bar hive, in that the beekeeper is working from the back of the hive, well away from the guards, and is not tearing apart boxes of honeycomb. This results in a less stressful inspection process for both bees and beekeeper.

The key difference between the Slovenian AZ hive and most other hive types is that this design is not meant to be left on its own in the middle of a field. With this system, the bees have their own custom-built house, with the hive working like a cabinet set into the wall. In the harsh mountains of Slovenia this means that the bees are basically living inside a house, with only the foragers leaving through the entrance of the hive to the outdoors. As a result, the beekeeper can check the hives year-round from the comfort of the hut. Under extreme Australian conditions, this also protects against heatwaves, which are the major killer and stressor for bee colonies. The obvious disadvantage is that the hive shed cannot be moved — although some ingenious beekeepers in Slovenia have got round this by building the hive sheds onto trailers, like a jumbo campervan for bees!

Harvesting from the AZ hive is similar to harvesting from a framed Langstroth hive.

Frames of capped honey need to be removed, and the caps are then cut off and the honey spun out, with the frames returned to the hive to be refilled. It is also possible to split a hive by removing frames of brood and honey to form a new colony, so it also has many of the advantages of a Langstroth or Warre hive. The main drawback is simply finding one! AZ hives are not much used outside of Europe, and there are very limited suppliers, although this is changing as more beekeepers realize the benefits of this approach. With the increasing prevalence of heatwaves, drought and damaging storms in Australia, housing bees in their own purpose-built shed is a tried and tested approach to adapting to extremes of heat and cold.

The big disadvantage that I discovered when working the AZ hive at the Embassy of Slovenia in Canberra was that the frames are not standard sizes — they are a little wider and a little shorter than a standard Langstroth frame that we most commonly use here in Australia. I found this out at the worst possible time: when I was about to extract honey from a hive that was very full and about to run out of space! Try as I might, I simply could not get the frames to fit into the extractor I'd brought along. The solution was to treat it like a top-bar or Warre hive, cutting out the honeycomb and crushing the honey out. I have since found out that the versions of the AZ hive that are being sold now all take standard Langstroth frames, so this will not be a problem in future.

THE TRIED AND TESTED STANDARD: THE LANGSTROTH HIVE

- Standard hive used all over the world.
- Spare parts easy to get, and there are good support instructional videos.
- Can involve heavy lifting — up to 25 kilograms or more in a high nectar flow.
- Excellent hive for splitting, joining and moving bee colonies around!

The discovery of a concept known as 'bee space' was one of the most fundamental discoveries in modern apiculture, and this allowed the development of the first truly modern hive by Reverend Lorenzo Langstroth in 1852. Bee space is the space that bees leave within their

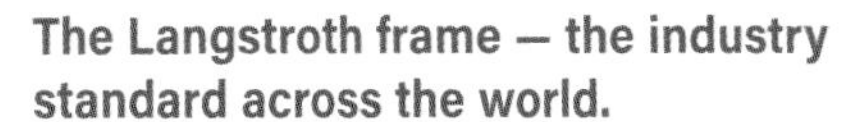

The Langstroth frame — the industry standard across the world.

The Langstroth hive manages space by having multiple boxes stacked on top of each other.

hives to allow for the easy passage of workers. Gaps in hives must be no smaller than 6 millimetres and no larger than 9 millimetres. Smaller gaps will be sealed with propolis. Gaps larger than 9 millimetres will be filled with honeycomb, so there is a maximum design tolerance of 3 millimetres that applies to all internal spaces the bees can reach.

The development of the Langstroth hive, as it is now known, was the greatest advance in beekeeping since skeps were invented, and basically created the modern beekeeping industry as we know it. It revolutionized beekeeping at the time, as it had removable frames allowing colonies to be moved, split to form new hives, and joined to strengthen weaker ones. Most importantly, the bees no longer needed to be hurt or killed to harvest honey.

The original Langstroth hives were made from wood, and were often dipped in boiled beeswax to preserve them. You can still buy these hives today, but you also have the choice of plastic and, my personal favourite, foam hives. Boxes of eight to ten frames are stacked on top of each other, with one or two bottom boxes forming the brood nest. 'Supers' for honey production are then stacked above this. The frames are fully enclosed and often have foundation wax placed into them to keep the honeycomb straight for ease of harvesting.

The brood boxes and honey supers are separated by a queen excluder. This is a slotted screen made of metal or plastic containing holes that only the worker bees can squeeze through. This ensures the boxes above this screen only contain honey, as the queen cannot reach them to lay eggs in the honeycomb. For commercial beekeepers, this is an important timesaver — the queen and brood are always below the excluder screen. For hobbyists and smaller scale commercial keepers, it is less critical as the bees will normally fill the uppermost box with honey first, but knowing that the queen is in the lower boxes can make inspections easier.

The biggest advantage of the Langstroth system is mobility. The fully enclosed frames and wired wax foundation make them resistant to the shaking of long truck journeys, and the entrances of the hives are designed to make it easier to close in the bees in the evening for overnight moves to new nectar sources.

For beginner beekeepers, there is also a wealth of knowledge and experience on how to manage Langstroth hives, and they are the first hive for many people (myself included). I use the foam versions for orchard pollination due to their light weight and superior insulation. As an added bonus, many foam hive designs are made to be fully recyclable, meaning less waste.

GOING WITH THE FLOW: GETTING THE MOST OUT OF THE FLOW HIVE

- Turns honey harvests into a fun spectacle.
- Looks beautiful, works really well and a nice way to keep bees.
- Spare parts easy to get, and good support instructional videos.
- Basically a standard Langstroth hive with a fancy (and clever) honey super added to the top of the hive.

Like a few other experienced beekeepers, I was initially dubious about the Flow Hive. It seemed just a little too good to be true, promising a simple, low-stress way of extracting honey. While the 'old guard' were initially opposed, I love trying out new hive designs and

The Flow Hive allows you to harvest gently without disturbing the bees.

Flow Hive frames partially filled with honey — not quite ready to harvest.

wanted to reserve judgment. I took part in the original crowdfunding and received one of the first production units, which was put into service in the Parliament apiary. Through public events like the UN World Bee Day, I got to meet and came to know the Flow Hive inventors, and found them to be lovely, passionate beekeepers. Their invention had grown out of a love for bees and a dissatisfaction with how disruptive and upsetting standard harvesting was for the colonies.

Fundamentally the Flow Hive is a standard Langstroth hive with a modified honey 'super' that allows the honey to be drained out of the hive without opening or otherwise disturbing the colony. That is basically it; all other management (inspections, splitting, swarm control) is the same as for the standard Langstroth hive.

As with any hive design, there are a few tricks I have picked up to get the most out of this design, and there are two areas where beginners seem to struggle:

Getting bees onto the Flow frames

The Flow frames are a food-safe plastic that is shaped like honeycomb, which the bees build on and complete to store honey — but they sometimes don't immediately start working them. Three tricks seem to have worked for me:

- **Obtain a block of pure beeswax and rub it onto the frames. This puts both the smell of a hive through the frames and gives the bees some raw material to start coating the Flow frames with their wax. You only need a little — I often use the side of a beeswax candle.**

- Spray sugar water onto the frames just before you put them in. This makes the workers move into the frames as they drink the sugar water, and they will immediately start cleaning the cells and working them as if they were their own.
- Add the Flow frames to a strong colony in the middle of a good nectar flow. The bees will be less choosy about their storage solution and will start to work the Flow frames more readily.

Always use a queen excluder

The Flow Hive is the one system where a queen excluder screen is essential. If the queen starts laying brood inside the Flow frames, the cocoons contain silk as well as wax, adding sticky cloth to a mechanical system. This is very bad news, and hard to fix. Always have the Flow frames sitting above the queen excluder to prevent this from happening!

Add a second brood box

A stronger colony means more production, and giving the queen a larger space to create the workforce seems to really work with this hive design. This means buying an additional brood box with frames, building a larger, stronger colony, before adding on the Flow super. The advantage is faster build-up of the colony and additional honey storage, with the second brood box acting as a reserve honey store to help the colony get through winter or a tough season.

Filling Parliament honey gift jars straight from the hive.

In the end, I was glad that I had reserved judgment, and I have found that a lot of the fears around the new flush of beekeepers entering the hobby did not eventuate. People loved their bees, undertook inspections properly and generally were much like me when I was starting out: making a few mistakes, but generally trying to do the right thing as they learned.

Over time I have become an unabashed fan of the Flow Hive. It popularized urban beekeeping, bringing a whole new generation of beekeepers into the fold. The company has showcased local precision manufacturing, proving that the public is prepared to pay for quality products. They have tried lots of new things to connect with their customers, from standard things like T-shirts to innovative projects like making native bee hotels from their waste products and offcuts.

There are a few versions of the Flow Hive available now, including the original version, the Flow Hive 2 and 2+ with some added bells and whistles, and what has become one of my favourite versions, the Flow Hive Hybrid. The hybrid version has both Flow frames as well as a set of foundationless frames that build sheets of pure honeycomb. I first came across the hybrid version when the Italian Embassy purchased some and asked me to help manage them. This combination allows you to take out honey quickly and easily, straight into jars. It also allows you to cut out chunks of raw honeycomb, which is one of my favourite ways to present honey. For an embassy that wanted to occasionally put honey into jars, but also wanted to present cut honeycomb on cheese platters, this was ideal.

There is a lot that all businesses can learn from the Flow Hive, particularly in terms of how to respond to criticism. The original 'plan' was to sell mostly to established beekeepers, but instead their launch video went viral, inspiring literally thousands of people to take up recreational beekeeping. It was a global phenomenon and was life-changing for the inventors. In one afternoon, they went from a couple of tinkerers in a shed out the back of Byron Bay to heading a multinational manufacturing operation.

With this flood of 'newbees' came a flood of concern from existing beekeepers. Would these new keepers properly care for their hives? Or would they neglect the bees, failing to inspect the hive and just treating it like a toy? The response from the inventors was positive and inclusive, helping to build an online support community. Some of the key measures were:

- **creating high-quality online instructional videos to guide new beekeepers on the basics of hive inspections, disease control and honeybee biology**

Honey being refilled into harvested Flow Hive frames.

- working with beekeeping clubs to provide teaching materials and discounted demonstration hives
- being available for conferences, trade shows and open days, and
- gathering tips, suggestions and design hacks and incorporating these into new versions.

There remains some criticism from the 'old guard' of beekeepers, and one of the main challenges new beekeepers have to face is a barrage of (often conflicting) advice. Flow Hive users get a double helping of this. But these measures have largely been a success. In my experience, the Flow Hive 'newbees' are diligent, enthusiastic beekeepers who love and care for their colonies. Like the rest of us, the bees have become their main fascination, with the honey a welcome by-product.

BELLS AND WHISTLES: HANDY GADGETS FOR YOUR BEEHIVE

While not essential, the modern age has furnished us with a wide variety of gadgets and toys, and it seems a shame for the bees to miss out!

In-hive monitors

One of the most wonderful things with the modern age is something called the internet of things. Put simply, this is taking inanimate objects and putting them online, where we can see them, hear through them and even make them move remotely. The first internet of things device was a Coke machine at Carnegie Melon University that was connected to the internet so that people could check if it was stocked, and if the cans were cold. But what if we need to check if our hive is full, or if it is about to swarm? Recent advances in sensors and artificial intelligence have made it possible to monitor our hives remotely, and even tell us how the bees are feeling at that moment.

My personal introduction to hive monitors came through an international research group called the Open Source Beehives Project. This

An early prototype hive monitor.

Top-bar parts cut with a CNC router.

Readout of temperature and humidity from an in-hive monitor, delivered to your smartphone.

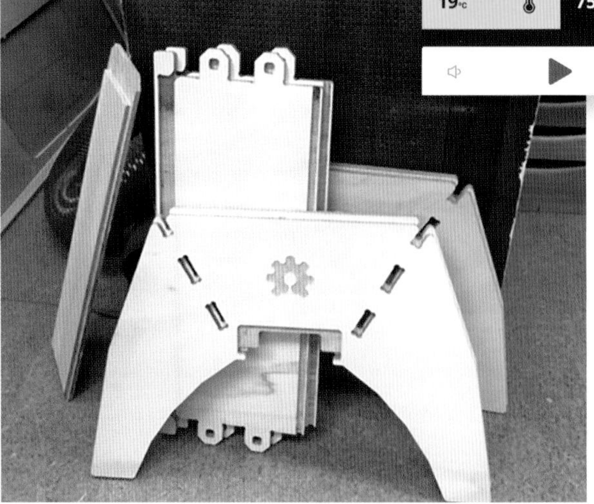

was a group of programmers, data scientists and beekeepers who initially focused on creating low-cost beehive designs that could be produced with a Computer Numeric Controlled (CNC) router, which cuts wood into precise shapes. This was essentially creating a hive like a giant jigsaw that could be cut out of a single sheet of plywood with minimal wastage, then slotted together with just wedges and gravity. No glue, screws or nails required!

The next phase of the project was developing online monitors that could 'listen in' to the hum of the colony and determine their health. We already knew that bees used sound to communicate, and I was one of the beekeepers who collected high-definition audio to train a learning algorithm to relate these sounds to hive health indicators. Eventually the system was able to detect swarming (which is really loud), pre-swarming and queenless hives, and distinguish these from a normal healthy state.

Once the system was developed, the focus was on field-testing units to place monitors inside hives in a way that did not disrupt or annoy the colony. This was a bit harder than it sounds, as bees can 'see' electricity via a network of fine hairs on their bodies. In nature they use these to detect the slight positive charge of a heavily laden flower, but when electronics are inside their hive the electrical components annoy them. The bees chew on the wires and encase the circuits in propolis, damaging them and gumming things up. The solution was low-power circuits that were less of an annoyance and secure housings that the bees couldn't damage.

The system that I helped to develop, the Buzz Box, has unfortunately gone out of business, but a few units were installed in the Parliament apiary for several seasons, and seemed to work quite well. In the most recent season it helped me determine when the colony had 'woken up' early for spring, allowing me to get ready to inspect to head off swarming. While this system might have fallen by the wayside, it is far from the only system available these days, and there are several more in development. As more systems come online, more ideas and insights will be generated and more data on bees will become available. While there are only a few thousand hive monitors in use at the moment, I expect this will become an increasingly normal aspect of modern beekeeping.

The main downside to these hive monitors is collecting and transmitting the data. In the Parliament apiary we had to set up a dedicated wireless data link, and for these systems to work something similar has to be in place. This really limits things in remote areas, but in urban areas there is often a power point close by, and most buildings have multiple wireless networks that could be used.

Hive scales

The holy grail of online monitoring of beehives is hive scales. These are incredibly useful, but also impossibly fiddly to set up and maintain. To date the only hive scales that I have seen used are owned by researchers with a lot of technical skill to maintain them, but I think it is only a matter of time before this becomes a more common addition to hives. Knowing the weight of the hive in real time tells us a lot about the state of the colony.

At the most basic level a hive scale helps you keep track of honey stores inside the hive, and can warn of potential starvation in winter and when there is enough honey to support a harvest. For native bees that provision cells, the net weight of the provisioned cells could also be used to determine population, but this would be prohibitively expensive for most native bee hotels. A simpler approach could be to simply record the weight of the empty hotel, and then periodically re-weigh it during the season.

Nectar flows

While the weight of honey in a hive is one metric that can tell you when to harvest, it is also important to take honey on a nectar flow, when there is fresh food coming into the hive, and the bees will be less stressed about honey being removed. A steadily rising weight allows you to pick that the colony has found a source of resources, and this can also tell something about the landscape. It is possible to determine the source of honey based on the pollen and sometimes also the flavour, and this can be matched with the increase in hive weight to determine what the best producing flowers are in your local area.

Hive population

One of the clever calculations that hive scales can do is compare the minimum weight of the hive just after the foragers have left to start work against the maximum weight early that morning, before the foragers left. The drones tend to not leave until later in the afternoon (lazy bastards) so dividing this weight by the average weight of individual bees gives you the number of foraging workers in the hive. This is incredibly useful, giving a good measure of hive strength, and can also warn beekeepers if there is a sudden drop in the number of foragers in the hive.

Unfortunately, getting a robust hive scale set up that is accurate and low maintenance has proven to be elusive. However, the fact that this information is so incredibly useful to hive management means that people are still trying, and someone will eventually crack it

— probably in a way that makes us all smack our heads and say, 'Why didn't I think of that?' At the moment there are a few hive scales available, either as 'build it yourself' kits or as add-ons to hive monitoring systems.

Theft detection

One of the biggest downsides to having a hive that is visible to the public is the risk of damage and theft. Random acts of vandalism are hard to predict and guard against, especially with a highly visible hive, but making the hive physically hard to reach helps.

Theft is harder to predict and stop. It might seem strange for most people to think of hives as something that would be stolen, but it is a very real issue. Every year thousands of hives are stolen worldwide, and the reason is basically the same: money. The hive itself can cost anywhere from several hundred dollars to nearly $1000 for well-made top-bar and Flow Hives, and that is before you even get to the bees. A fully established bee colony is worth upwards of $200 in most places, and the honey can be worth several hundred more.

Most theft on a large scale is unfortunately other beekeepers, often to service pollination contracts. With pollination services running to several hundred dollars per hive, there is a lot of money at stake, and unfortunately every year large-scale 'bee rustlers' make off with hives from fields. In urban areas thefts will be smaller in scale, and often involve high-value hives that are visible from public areas.

I have one sneaky method for detecting theft that has worked so far: my hives have small electronic trackers hidden inside them. I use Bluetooth trackers, which are detectable by users with the same app on their phone when they pass

Electronic hive trackers are cheap and hard for thieves to detect, especially inside the hive.

Hives for Humanity in Vancouver, Canada has well secured hive sites.

close to the hive. Only I can see the location 'ping', which makes the trackers hard to detect. I have them built into the sides of the hive or inside the frames, which the bees then cover in honeycomb, basically making them completely invisible. The one weakness of the system is that another user with the app has to pass by, but the system I use (Tile) has over 1000 users in my area, so I think it might be a bit tricky to sneak a hive past all of them.

The other tried and tested method is to make the hives physically difficult to reach. My rooftop hives are on top of access-controlled buildings, and the Parliament hives are surrounded by some of the best security in the world. This is one of the big advantages of rooftop hives, in that you get to 'piggyback' off the building security, which is normally many times above what is necessary to deter direct theft.

When I visited one of the Hives for Humanity sites in Vancouver, Canada, I was impressed at how open and accessible their sites were. They were right in the middle of the city, providing a showcase of the connection between bees and landscapes, but with this came physical security for the hives. One site that I visited in Canada was in a park up a series of steps that restricted vehicle access, and the hives themselves were behind a sturdy metal fence. Pinching these hives would require a lot of time and effort in a very public place, and it probably wouldn't be worth the effort for most potential thieves.

Building a secure apiary needs a balance between public display and security that includes physical barriers and theft detection systems, but the benefit is peace of mind that your bees are secure, or at least that you can recover them if they are stolen.

7

Hive management

Connecting with your bees

When you tell people that you are a beekeeper, most folks immediately think of honey, and some people start their beekeeping journey with this as their initial interest. It is undeniably fascinating to take the floral bouquet of the landscape and collect it in a magical, concentrated form that is unique to where you live. What could be more interesting? The reality is that the bees themselves are endlessly fascinating, and spending time with them is often far more rewarding than any momentary sweetness (though that is still lovely).

Once you move into the space of being a more public beekeeper, with bees that are in a park, community garden, or perhaps a rooftop you run tours to, there is an even greater focus on good beekeeping practice.

OBSERVATION

While there is no substitute for opening the hive and checking brood for disease control, external inspections are an important part of checking on your bees. As an added bonus, it is fascinating and helps to connect you with both the colony and the wider landscape.

Observation of the hive and the surrounding landscape is the cornerstone of good beekeeping. Deep interest and reflection are what distinguishes someone who happens to have a beehive in their yard from a true beekeeper.

There is a beautiful, almost meditative calmness that comes from sitting next to a beehive, watching the bees work. You are placing yourself on their doorstep, and in the same way as you can often pick up on the mood of a household from the outside, you can discern a lot by watching what is going on and by watching the foragers leaving and returning.

HOW CHILL IS THEIR BUZZ?

When I visit the Parliament hives in their semi-public area, sitting next to the entrances for a few minutes is more than just a pleasant interlude in my day (although it is wonderful). It is a critically important test of the hives' overall mood and tolerance for the outside world.

Will the colony tolerate your presence if you just sit quietly next to them and watch them work? This basic question tells us a surprising amount about a hive, and both honeybees and native stingless bees have a set of 'tells' about their mood, some subtle, some less so.

I was once showing a TV celebrity the Parliament honeybee hives. He was initially terrified of being near the bees — all his life he had been taught that they were just itching for a chance to sting us. The hives were busy, with lots of nectar and pollen being carted in, and a traffic jam at the entrance. However, the gentleness of the colonies shone through, and he was no idiot. After a few minutes, trepidation gave way to fascination, and he even had returning workers gently landing on him for a rest before they returned to the hive. This is how the colony should be for you: a busy hustle of workers coming and going, happy for you to be close to them and even occasionally resting on you on their way back.

His initial fear was not completely unfounded — some bee colonies become jerks, liable to mount a vigorous and disproportionate defence

Calm bees hanging out on the front of the hive.

A queen with cranky genetics (now much quieter).

at the slightest intrusion into their territory. This defensiveness is a really important tell: not only is the colony a hassle to be around, it is potentially a threat to neighbours and even yourself when inspecting. It also makes beekeeping a chore to have a continually cranky colony.

Cranky hives can also be an indication that something is going wrong with the colony. A hive that progressively becomes less and less pleasant can often be a sign that a gentle queen has been replaced by a feral one. The colony will change temperament as the old, gentle genetics are replaced with new ones (and vice versa). A sudden change in demeanour is more worrying — is the hive noticeably louder and crankier? This is the classic sign that the queen has died suddenly and the colony is desperately trying to raise a new one.

It might also be under attack. European wasps, ants and even some animals will harass a colony and turn it defensive. Perfectly reasonable — if jerks were continually trying to sneak into your house to steal your stuff, and even your kids, you would be pretty pissed off too. Watch for evidence of intruders and check Chapters 8 and 9 on pests and diseases for what to do if you see some unruly interlopers.

Probably the strangest and most fascinating reason for a cranky hive is also the least concerning: the smell of cut grass. That lovely cut-grass smell that we enjoy is actually the grass screaming in pain. When cut, it sends out a chemical alarm signal that is blown on the wind to surrounding grasses, which start making bitter compounds in their leaves (this is one of the reasons herds of cattle graze into the wind). Bees are intimately connected to the plants that surround them and they will pick up on this alarm signal and act as if there is a threat. We still do not really understand why they pick up on the alarms from grazed grass, but you really don't want to open the hive or hang out next to it after mowing!

SMELL

I sometimes feel shortchanged to be born a human. Despite being the overwhelmingly dominant species on the planet, we are born with such a rudimentary set of senses. The human nose works perfectly well, but bees keep their noses on their two antennae, meaning they smell in 3D!

The smell of a healthy honeybee hive at work is a beautiful thing. As nectar is converted into honey, the workers fan their wings over the honeycomb to evaporate off excess water, and this creates a haze of delicious honey scent that surrounds the hive. A strong smell of honey as you approach means that nectar is flowing into the colony, and this is a smell that makes every beekeeper happy. It does wonders for the mood of the hive as well.

Conversely, a lack of honey smell can signal that food is scarce. Worst of all is a rotten or fermenting smell. This often means disease — the major diseases of American foulbrood and European foulbrood get their name from the rotting smell of the dead brood that they create. See Chapters 8 and 9 for information on pests and diseases.

NOISE

The hum of the hive has fascinated people since we first started writing about bees; their buzz is synonymous with nature. The hum that a bee makes as she works is a beautiful part of the natural soundscape of a garden, but en masse it can be an amazing, visceral experience. I once visited a Lavender farm in full bloom, and as I parked the car, I became aware of the buzzing hum in the air. As soon as I opened the door it was overwhelming — the sound of literally millions of bees, working the flowers. One of those visceral experiences that really stuns you with its power, and stays with you. When a hive is really working hard you will experience this on a small scale, with the sound carrying over to trees and flowers in full bloom, as if we are adding a nature soundtrack to our cities.

Bringing it back to the smaller scale of the individual hive, the sound of the colony is an important indicator of the state they are in. A happy hive has a constant, gentle hum that is almost hypnotic. This stays at the same basic pitch whether they are working hard (when it simply gets louder) or not. Just sitting next to your hive for a few minutes a day will help you tune in to the sound of your colony.

Sudden changes to the pitch of the hive are worth noting. One of the classic 'tells' that the colony is replacing their queen is a hive that roars with activity, even in the evening when they would normally be quieter. Changes can be benign, but sometimes not. If in doubt, it is time to open the hive and check that everything is okay.

During an inspection, the pitch will normally stay similar to the normal working hive sound, but will suddenly change when they are losing patience with you. Experienced beekeepers often talk about the hive 'growling' at them. This is basically the same as a dog growling at you; it's the first concrete sign that they have had enough for today and that it is time to back off and give them a rest. Use your ears; it will save you from a lot of stings!

ENTRANCE ACTIVITY

One of the unexpected things that came up when I started establishing beehives at work and then later at Parliament was how much people love to watch bees. Watching the comings and goings of the bees as they work is one of the delightful, mesmerizing parts of having a bee colony in your garden or workplace.

Not just honeybees either — the native stingless bees have a firm following at Parliament as well, and the fact that people can safely approach quite closely means they have become social media stars. There just seems to be something therapeutic about watching bees work, their activities

Watching bees work is surprisingly calming and fascinating.

changing throughout the day. There is the swirl of bees orienting themselves first thing in the morning, flying in ever-increasing circles as new foragers lock onto 'home'. During the day there are rushes and lulls as bees foraging at different ranges return to drop off their goods. One of my favourite things is watching the mad rush when there is a change in the weather to rain — the foragers race home to beat the storm, scrambling to get inside.

Just observing can tell you so much about the hive, and I became a lot more analytical watching the bees as I gained experience. Here are some of the things to look for:

Numbers: The most basic thing to observe is how many bees are coming and going? Has it gone up or down recently? A sudden drop in activity without cooler weather can signal trouble. Many diseases will slowly sap the energy of the colony, making the foragers look listless.

Landing: Are the bees landing perfectly or do they sort of crash-land and tumble into the hive? The latter is better, as it signals that the foragers are heavily laden with nectar and/or pollen, which can be as much as their own weight.

Pollen: Both honeybees and native bees use pollen as a critical resource, and its colourful nature means that it really stands out on the foragers. Honeybees and native stingless bees carry pollen on their legs like fancy pantaloons, whereas some stingless bees carry it on their belly, but it still stands out. The cool thing is that plants have wildly different-coloured pollen, so you can tell what your bees are foraging on.

Beelining: Honeybees in particular will send each other co-ordinates and foragers will leave the hive and fly straight to that source. This is where

Bees clustering at the entrance when the hive is overheated is known as 'bearding'.

There is no substitute for regular inspections of your hives.

the term 'making a beeline' for something comes from, and you can get a bearing towards the flowers and then go for a walk to see if you can find what they are foraging on. Are they working one large source or heading off in multiple directions?

Dead bees: There are specific undertaker bees that take dead bees around 1 to 2 metres away from the hive entrance. Seeing piles of bees closer than this can mean that they are getting overwhelmed and taking shortcuts, which can signal pesticide impact or disease.

One really important thing to note: just observing entrance activity is not a suitable substitute for regular inspections. Some of the original texts for the Warre hive and other systems talked about using entrance observation as a significant way to monitor hive condition, but this was before brood diseases like American foulbrood became prevalent. Observation tells you about the foraging success and temperament of the hive, but cannot substitute for proper inspections of the colony. It's still a lot of fun, though!

OPENING THE HIVE

While you can tell a lot from what is going on outside the hive, there is no substitute for actually opening the hive and inspecting the brood for disease. This is why there is such a strong requirement for beehive designs to include removable frames that allow for direct inspection of the brood. Even a top-bar design with minimal control of comb-building meets this requirement, allowing slices of honeycomb to be taken out and checked.

Beginning beekeepers often focus on the honey stores, but to be honest most experienced beekeepers move straight past it, taking just a quick note of honey by lifting it to check roughly how much there is and whether it has been capped (sealed into honeycomb). It is important for the colony to have enough stored food, especially going into winter, but mostly it is not the main point of an inspection.

The key objective of an inspection is to determine the health of the workforce rather than the fruits of their labours. You don't even need to see the queen, as long as you can see sufficient evidence that she is present and laying healthy workers to replenish the colony.

Each inspection should meet some basic objectives. In order of importance, these are:

Queen right: Is the queen present and laying? The simplest way to check this is to actually see the queen, but eggs don't move and are the next best thing — they look like tiny grains of rice in the bottom of empty brood cells.

Brood health: Checking both the capped and uncapped brood is the next most important thing — it should be in smooth sheets with minimal gaps. We are specifically checking for sunken, greasy-looking capped brood, or yellowing, mouldy-looking uncapped brood.

Swarm or supersedure queen cells: This is where the colony is looking to create new queens to replace a queen that is either leaving with a swarm or has begun to fail.

Equipment condition: Look for any signs of mould, damage or black-looking brood comb that needs to be replaced, and remove burr comb that isn't being built out straight.

These are the things you should be inspecting for with every inspection, but the additional things that you search for depend on the time of year.

PLANNING INSPECTIONS

To be successful in almost anything you really gotta have a plan! Beekeeping is no different, and when you are dealing with a colony of sometimes cantankerous insects, this is doubly so. To be fair to honeybees, the process of pulling apart the hive for an inspection looks, sounds and feels a lot like a bear attack might. On top of this, exposing the brood to cold air can kill them, threatening the whole colony, and the queen could get lost or damaged, so they have a legitimate reason for limited patience!

The trick to inspections is to have a clear idea of what you absolutely need to achieve, what is nice to check on and what is irrelevant. Interestingly, for most of the year honey falls into the last category. Unless honey reserves are critically low, they are the thing you move past quickly. What really matters is the health of your workforce: are they replenishing themselves and are they disease free? These are much more important questions than how much honey is on hand — look after the health of the colony and you will both have plenty. As an established beekeeper, you should have an inspection regime already established for your home hives and this can be adapted to any community or rooftop hives. Here in Canberra, I inspect around once a fortnight in spring, once a month in summer/autumn and just do brief external checks through winter. Other areas of Australia will allow for year-round inspections, though, so adapt to your local area.

KEEPING RECORDS

This is now a legal requirement in most jurisdictions, but it is also a useful exercise in keeping tabs on the long-terms trends in your apiary. The question that often comes up is what is 'normal' for your area? I used to think that I would remember each season perfectly well, but I was wrong; reading through my notes from inspections I would always pick up some additional small details that I had forgotten. Over time, reading and reviewing what you did in past years hones your instincts as a beekeeper, and is also something you can share and discuss with mentors and colleagues.

In the training apiary that I help run with a group of other beekeepers, record-keeping is also essential to allow us to pass detailed notes to each other when we can't physically catch up. While we still send emails to note key tasks, the detailed notes taken at each hive in our electronic record system means we are prompted to check for specific things, and we don't miss recording small details.

WHEN TO SPLIT, WHEN TO JOIN: ESSENTIAL SKILLS FOR THE URBAN BEEKEEPER

Beyond the basic management of a single hive, there is a range of techniques that all beekeepers need to know how to do, but these become more critical for urban beekeepers. Being able to split, join and manipulate colonies is an essential set of techniques to re-home a captured swarm and prevent new ones. While this is an unnatural intervention in the lives of the bees, when done properly it can save a weak hive and prevent swarms from escaping to bother the environment.

Joining colonies (Warre, Flow and Langstroth hives)

When the Slovak ambassador wanted to create a beehive to populate a traditional cabinet-style hive that he was about to take delivery of, we were in the middle of an intense swarm

Three separate colonies stacked in nucleus boxes, about to be joined into one.

Hives all joined together in the beautiful Slovenian hive enclosure.

season. All the collectors were run off their feet, and we had literally dozens of new colonies to deal with. It was a great opportunity to merge several swarms to instantly create a cohesive (and much stronger) one.

A trio of captured swarms had been delivered in nucleus boxes, and one Sunday I turned up at the embassy with two extra hive boxes, extra frames and a newspaper. Less than an hour later three swarms had become one large colony, separated by thin sheets of newspaper that the bees were rapidly chewing through to merge with their former neighbours. It was fun explaining this to the ambassador, who had never kept bees — we tend to forget how absolutely insane merging colonies sounds.

Joining colonies is a fantastic option when you have several weak colonies or a spare swarm that you don't know what to do with. The additional workforce significantly strengthens the receiving hive and is a good way to save a weaker colony from starvation in autumn. Provided that both hives have been inspected and confirmed to be disease-free, the risk is very low, as long as the join is done gradually enough to give the colonies time to meet gently.

Like a lovable, buzzy Frankenstein monster, merged colonies are a great way to create a pollination powerhouse, and such a merge takes advantage of bees' adaptability. The stacked boxes of the Langstroth/Flow hives make it simple to build the strength of an existing hive by 'papering on' a swarm to an existing colony to form a single large colony. The process is made easier by the top cloth that covers the Warre colony — you simply remove the top cloth, replace it with a sheet of newspaper that has had some small holes poked in it, then place the second colony on top. The two colonies will gradually merge to form a single large colony.

This results in an instant boost to the workforce and allows the bees to rapidly take advantage of pollen and nectar flows. It's a good idea to carefully check, and give them plenty of extra storage space if you have a good nectar flow coming in, as they will really go for it. You will generally lose one of the queens in this process, so if there is a 'nice' queen that you would like to keep, make sure you remove the one you don't want first!

Joining colonies (top-bar hives)

The internal arrangement of these hives makes joining colonies quite tricky, as it is more difficult to keep the colonies apart. You also need to have a top-bar hive design that has enough space (and ideally a side entrance) to allow you to have two colonies side-by-side. One advantage is that the table-like top of the hive makes it easy to place a nucleus on the top of the bars and gently lift the colony to be joined into the empty space next to the main colony.

The basic principle is the same: once the two colonies are in place next to each other, use paper or some other medium that the bees can easily chew through to slow down the introduction. I have seen beekeepers tape or use propolis to stick newspaper into a dividing screen inside top-bars, and some follower boards (used to confine the living space of the colony) have a handy hole drilled in them that is normally sealed with a cork. Remove the cork and insert a

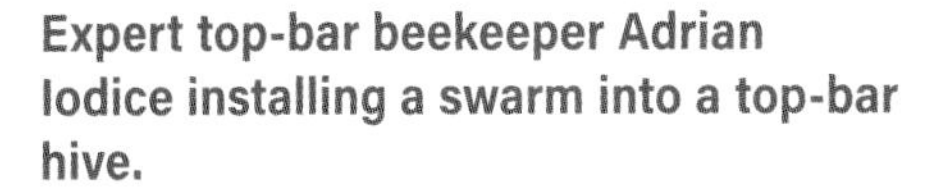

Expert top-bar beekeeper Adrian Iodice installing a swarm into a top-bar hive.

marshmallow in this hole, and you have a delicious way for the bees to meet their new friends.

It is important to be as smooth in your movements as possible and use minimal smoke during the process; you need as much of the colony as possible on the combs, as you have less control over the bees in the top-bar system. Just enough smoke to get them up onto the combs, but not so much that they start running around on the bottom board is what you are aiming for. Once they are all in, the remaining bees are gently brushed into the hive before you close up by replacing the top bars.

Splitting a hive using your nucleus box

The most common use of a nucleus hive is to split a hive, creating two colonies from one. This is not just done to increase your colonies; it is one of the most effective ways to control swarming. Uncontrolled swarms are a threat to the environment, and also mean half your workforce is heading out the door! The split consists of five to six frames of honeycomb from the 'mother' hive, plus their queen, being moved over to the nucleus hive to create the new colony. Once the main colony realizes that their queen is gone, they will raise a new one.

In terms of swarm control, this counts as a successful swarm as far as the colony is concerned, and it alleviates the impulse to swarm again … at least for a while. My usual trigger for splitting a colony during spring inspections is the presence of swarm cells. These are queen cells placed along the bottom edge of the frames, and signal that the colony has started the swarming process. Some beekeepers remove these, but this is a mistake. The colony has made up its mind by this stage, and splitting it into two is the only really effective treatment.

The important point here is to keep the process as close to the natural one as possible; in the same way that the swarm would take a cross-section of the colony, you should try to take a few frames with honey, pollen, uncapped brood and capped brood. A split needs to be a mini-hive that can survive on its own, and must have a few common components:

- **a queen (or enough eggs to make one)**
- **provisions (at least one frame of honey, ideally pollen as well)**
- **brood and bees (ideally multiple ages).**

It is also worth marking the queen as you transfer her over. Quite often a swarming colony

Splitting a colony into a nucleus hive is a highly effective swarm control method.

will replace the queen in the months afterwards, and marking her will let you keep track of this.

There are a few different ways to do a split, and I tend to do these at different times of the year. The basic approach is the same, with variance in the details.

These are three basic types of splits that beekeepers commonly use:

Swarm prevention

Early in the season, I always have a nucleus box handy to split into if I see any signs that the colony is preparing to swarm. Swarm cells are generally pretty easy to spot: look for long structures that look a bit like unshelled peanuts stuck to the bottom of the frames. If I see swarm cells starting to form, I grab a nucleus box and do the split straight away. This split takes the old queen out on a frame with no swarm cells, and you then build the new nucleus colony around her with frames of honey and brood. This simulates a natural swarm and is really effective in getting the desire to swarm out of their system.

Critical element: You must remove all of the swarm cells from the frames going into the split, but leave some behind in the parent colony.

Walk-away split

This is the classic split used all over the world to build the number of colonies in your apiary. You build the split with frames of honey, brood and eggs, and then basically just walk away and let the bees sort out creating a new queen. This can be done fast, and if it doesn't work you can merge the (now queenless) colony back onto the parent hive. The big advantage of this approach is that you don't even need to find the queen. As long as both the split and the parent colony have frames with fresh eggs on them, the queenless one should raise a new queen.

If for some reason the queenless colony can't raise a queen, you can keep transferring frames with eggs into the colony, or simply admit defeat and join the split back onto the original hive. As a result, this is a low-risk way to increase your hive numbers.

Critical element: Make sure there is honey, brood and eggs in both the hive and the split so that they can easily raise a new queen. You don't actually have to find the queen — just make sure that both colonies have eggs so that the queenless one can raise a new queen.

Splitting a hive into a nucleus box at one of my rooftop apiaries.

Requeening split

The most common reason for me to do a late-season split is to requeen, especially if I am dealing with a really defensive colony. A defensive colony will sometimes not accept a queen if you just put her in with a few attendants, even if you have removed their current queen. They will just kill her and raise their own. There are other ways to slow down the introduction (like push cages; see p. 160 for more information) but my preferred method is to build a solid posse around the new queen that can defend her against the hive you are trying to requeen. Making a split means that the new queen's pheromone can reach more of the new, smaller colony, increasing the likelihood that she is accepted.

The other reason I do this is weather. Queen breeders need to send off queens on a set schedule, but sometimes when they arrive in the post it's a terrible time to try opening a hive, with wind, rain or a day job getting in the way. This is especially the case on city rooftop apiaries, where there are also sometimes high winds to contend with.

Making up a queenless split once I know that the new queen is in the post and only a day or so away means the split will realize that they need a queen, and I can just pop the cage straight in as soon as she arrives (seriously, it only takes 30 seconds). I can then wait for a better day to find and remove the old queen, and then paper on the new queen with a colony surrounding her to defend her.

Critical element: You need to make sure that the queen you are trying to replace is not on the frames you move into the split. This is much easier than actually finding the queen in the larger hive, and you can do it once the mail tracking shows the new queen to be two or three days away to maximize acceptance.

Marking the new queen in this nucleus hive.

REQUEENING A DEFENSIVE HIVE

A 'cranky' queen with defensive genetics dispatched.

One of the things that really stresses new beekeepers out when requeening the hive is finding the old queen they want to replace, and dispatching her. Even as an experienced beekeeper, I can't always find her majesty during inspections, and it would stress me out too, except that I always have a nucleus with me when doing this, and use the requeening split method noted above. But sometimes you really need to fix a cranky colony that is causing problems.

You need to remove the old queen, and there are two ways to go about this. First, get an extra (ideally experienced) beekeeper and manually search through the original hive during the next inspection to find and remove the old queen. This is the only really effective method for top-bar hives and others with a horizontal arrangement of frames, and the best approach is to take at least three bars out, creating a 'light gap' in the sun that the queen will be reluctant to cross. You then systematically work through until you find her.

You can do it by yourself, but sometimes a queen is a 'runner' who is good at hiding; I even had a wild queen once who preferred to run around the edge of boxes, rather than staying on the combs during inspections. She had created quite a nice, productive colony, so I don't know what she was so worried about! A second pair of eyes is extremely handy, and at least the colony will have someone else to beat up.

Manual searching is fine, but sometimes you have to deal with a really cranky colony that jumps all over you as soon as you open the lid. This makes finding her majesty quite difficult, so you need something to get the numbers in her personal army down. In these situations, I use a trick that I picked up from American beekeepers dealing with the super-defensive Africanized bees. Also known as Killer bees, these varieties of honeybee are ferocious and can be quite dangerous. Luckily, there is a good method that never requires you to take off the suit or the gloves!

Here is how it works:

- Place queen excluders between every box (you can even cut them to size for Warre hives) and leave them in place for a few weeks. At the same time, make up a nucleus hive with the new queen you want to introduce and place it right next to the main hive.
- Set up a top and base that fits the hive boxes around 50 metres away (or more), and have a spare box handy next to the main hive.
- Check each box until you see brood. As soon as you do, pick up the whole box, sandwiched between the two queen excluders, and move it to the new top and base.
- Seal the top of the hive with the sheet of newspaper, and poke a few (small) holes in it with your hive tool, and then place the empty box on top.
- Transfer the nucleus colony into this top box above the newspaper. The bees will eat through the newspaper, and the two colonies will gradually merge. Don't forget some extra empty frames to fill out the box!
- A day later, go back to the box you moved and search through it. The foragers will have left and then gone home, meaning there will be a dramatically smaller number of bees that you have to go through to find the queen. Once the 'old' queen is out, you can paper this colony back into the original.

This method takes a while and has lots of steps, but it is extremely effective and is relatively simple to do, even in full gear. The best thing is that it allows you to fix a cranky hive with minimal disturbance, which will hopefully keep the bees a bit quieter as you complete the process!

Nucleus hive with a new queen set up next to the target hive.

8.

Keeping a healthy hive

First steps to disease prevention

There are relatively few things in beekeeping that are absolutely hard and fast, but managing pests and disease is one of them. As a beekeeper, you must absolutely be on top of this, and it is the primary reason that we inspect beehives. This is also why some older types of hives are no longer legal to use — hives such as long log hives and wicker skeps do not allow access to the brood for inspection. A disease could potentially fester in these hives until it breaks out and spreads to other colonies, a potential disaster in the making.

This is not just about honeybees, either. Many native bees are susceptible to diseases from honeybee colonies, so it is imperative that we don't allow these pathogens to spread into the broader environment.

When thinking about diseases from a management perspective, most of your focus should be on the vectors — the things that actually move the disease around in the landscape, allowing it to establish. Shut down these vectors, and even if there is a source of disease it can't move and will eventually die out. We saw this play out among humans on a major scale during the Covid-19 pandemic, where there were limited treatments and, early

Foraging bees spread pollen, but sometimes also pests and disease.

on, limited knowledge of transmission, so controlling movement of the disease became the main weapon in the fight against the pandemic. While the severe nature of the lockdowns was hard, it did illustrate how effective this approach can be.

There are four main vectors for disease in an apiary:

Equipment: This is the major threat of disease worldwide, with beekeepers moving infected gear between apiaries without properly cleaning or sterilizing gear. It is also the one we have the greatest control over.

Robbing: This is movement of infectious material by the bees themselves, and is again largely within the control of beekeepers, as the collapsing/weak hives that are robbed are most commonly in a managed apiary.

Flower-borne infection: Spread of disease spores through the environment is one of the major ways that keeping honeybees can impact on the wider environment, as well as infect other hives. Recognizing the signs of disease in our managed hives and then treating this to remove the source is a key thing that all beekeepers need to be doing.

Pest insects: Other insects can also spread disease, and infectious material has been found to be carried by several pests that commonly attack beehives. In addition to the undesirable impact of the pests themselves, these critters can leave behind some nasty surprises, which is all the more reason to aggressively control them!

APIARIES AS ISLANDS: USING THE BARRIER SYSTEM TO CONTROL DISEASE

It might have been Isaac Newton who said, 'We build too many walls, and not enough bridges', and while this is a good guide for our relationships, for biosecurity you really want walls. One of the best walls you can build between your bees and diseases is a management approach known as the barrier system.

The barrier system is one of the most effective innovations in biosecurity and is ideally suited to urban beekeeping. At its heart, this approach is about setting up your apiaries as 'islands', with you as the beekeeper moving between them, but nothing else. This is because most pests and diseases of bees cannot move very effectively; in most cases, it is beekeepers who move them between sites on infected gear. The solution is fairly simple: maintain a separate set of gear for each apiary, and maintain good record-keeping and sterilization for any gear that is removed for harvesting or maintenance.

For my urban rooftop and country pollination hives, this means a separate box of gear at each site. While this is mostly about disease control, one of the most critical factors that you will need to manage when starting out is time. Lugging gear around, especially to a rooftop, is a time-consuming hassle that you really don't need.

Even maintaining a separate set of gloves and hive tools for each apiary is a significant improvement in disease control, but I would strongly recommend setting up so that you can just walk up, grab all the gear you need and start working. If you think in terms of an hourly rate, your set-up costs will very quickly pay off!

The contents of your kit are dictated by the environment around the apiary. How remote is it? What am you going to need during a typical year? I tend to sort out my kits during winter, and this is where a thorough review of your notes from the previous season can be handy — check what you needed, and make sure that you stock the kit with these items before the hectic rush of spring rolls around.

APIARY KITS

Kit	Recommended contents
Basic kit 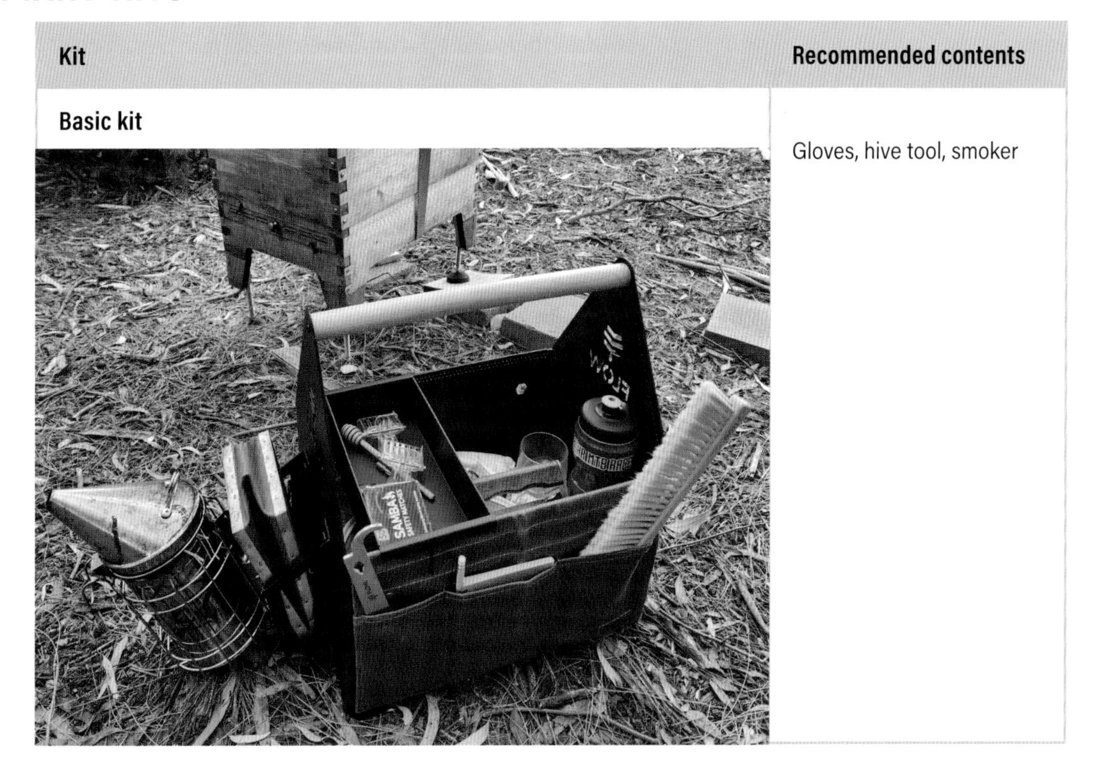	Gloves, hive tool, smoker

Rooftop hive kit 	Gloves, hive tool, smoker, spare suit, smoker fuel, spare straps, tape for emergency repairs Nucleus hive (with frames), swarm lures
Public area kit 	Gloves, hive tool, smoker, spare suit, smoker fuel, spare straps, tape for emergency repairs, first aid kit (including EpiPen), extra locks, sting relief Suits and disposable gloves for visitors

Keeping frames separate

Each Sentinel hive has its own kit, stored inside a false top section.

The real problem when it comes to pests and diseases is frames or bars of honeycomb, especially if they have residual honey on them. These are the natural habitat for many pests and diseases, and swapping frames among apiaries is one of the main ways that major diseases like foulbroods are spread. It is essential that the frames from each hive only go back into that hive. I use a system of labelled boxes when I am doing harvests, and you can also mark frames with the apiary name or colour to keep track. Seriously, do this and record it diligently and you will greatly reduce the risk of transferring disease and causing yourself all the treatment/replacement costs that come with it.

The Sentinel beehives used to detect diseases at airports and seaports as part of the National Bee Pest Surveillance Program take this to the next level. Each hive has a false top with an extra super added to each hive, with a complete set of diagnostic gear stored in the top. These hives need to be totally separated from all other hives due to the chemicals used in some of the detection tests, so they and all their frames and gear are marked with an unmistakable pink paint!

KEEPING HONEYBEES DISEASE-FREE

Once the barrier system is set up, you still need to inspect regularly for disease. There really is no substitute for getting into the brood box and checking the health of the colony. As noted in the inspections section on p. 142, for an outwardly healthy colony you should still regularly look at a sample of brood combs to check for eggs, uncapped brood and capped brood. If you suspect something is wrong, or if you have received notification of a disease report in the area, then it is time to get the heavy gear on and check each and every frame to confirm there are no symptoms. Even without a notification or signs of disease, during a full brood inspection you might also need to shake the adult bees off the combs so that you can see the entire frame clearly.

STARTING WITH A HEALTHY HIVE (AND A MENTOR!)

Put one hundred beekeepers in a room, and you will get about a thousand different ideas on how to keep bees (most of them mutually exclusive). Here is the thing, though: mostly these will be perfectly good ways to keep bees. Developing your own personal style and relationship with bees is one of the fun things about beekeeping. However, for a more public beekeeper, you need to set a good example.

These are the key things to look for and confirm during inspections:

Foraging behaviour: As you approach the hive, are they busy as, well, bees? Are they bringing in pollen or landing heavily as if laden with nectar? Are they actively guarding their entrance? Can you smell the sweet aroma of honey as they evaporate the moisture from nectar?

Brood pattern: The capped brood are the most common indicator of health, and this is where most beekeepers start. They should appear deep inside the hive in a section known as the 'brood cluster', which is basically a nursery for the colony. The frames containing the brood should be in a solid 'block' close together, and the cocoons of the baby bees should be densely packed as a solid sheet, with only a few

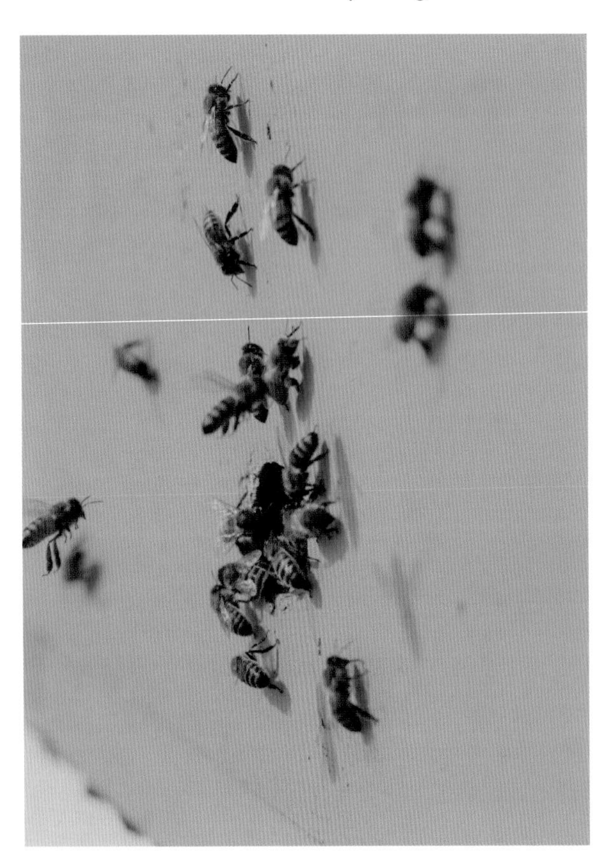

Entrance behaviour can provide some clues as to what the colony is up to.

What you want to see – a frame of healthy bee brood.

Frame with capped honey (top left), bee bread (orange/yellow cells) uncapped brood (white cells) and capped brood (brown cocoons, bottom centre).

gaps to allow the heater bees to maintain brood temperature.

Food and medicine stores: The frames should show a clear organization within the hive, with honey in the top of the hive, colourful bee bread above the brood and a decent amount of propolis sealing the hive and any gaps between frames.

Colony temperament: The colony should be tolerant of at least a brief inspection on a sunny day with no wind. The bees should be calm on the combs, meaning that they don't stop working when you pick up the frames or bars. You should also be able to calmly sit by the hive before you inspect and not be unduly hassled.

One you have set what is 'normal' in your mind by inspecting healthy hives with your mentor or a local beekeeping club, then you can start to look for the deviations from this that can indicate problems. This is the second part of where a good mentor is invaluable. I would often send photos to Carmen, my mentor, asking her, 'Is this normal, have you seen this before?' The best thing was that not only would she give me the answer, but she would also talk me through how she had made decisions on similar issues and provide pointers on what to look for in upcoming inspections.

Calm, healthy bees are essential for community gardens and education.

9.

What the hell is that?

The pests and diseases of honeybees

The diseases of honeybees are both horrifying and fascinating, and some of the pests that attack honeybees are downright ingenious.

AMERICAN FOULBROOD

It is hard not to be impressed with the resilience of American foulbrood (AFB) spores. Resistant to heat, cold, and even most chemicals, they can persist in infected honey, timber and soil for over 80 years, lying in wait for a chance to start a new infection. That is about as far as the admiration goes, though, as this is a truly horrible disease and in Australia is the major issue that we guard against.

AFB sets the perfect trap for the bees, the sticky spores lying in wait in the honey and on infected equipment for bees to accidentally pick up and feed to their young. Once the infection is established in the brood, the colony is doomed. The larvae incubate the bacteria,

Performing an alcohol wash test for Varroa mite in the Parliament apiary. (Credit: 5 Foot Photography)

which then kill the baby bees as they pupate, turning them into a smelly, sticky goo inside their cocoons that the adult bees desperately try to clean out. As they do so, the sticky nature of the dead bee ensures that the spores are spread throughout the hive, eventually making their way into the honey and pollen stores. These are fed to more larvae, causing more death until the colony starts to weaken and loses the ability to defend itself. This is often the trigger for other colonies to rob the dying hive, stealing the honey … which is now a trap filled with AFB spores. Once these are fed to the brood in the new colony, the process starts again.

AFB was not distinguished from European foulbrood until 1907, and by then it had spread widely. It is now found almost everywhere in the world where honeybees are found. However, the spores cannot move on their own, and bees can only fly so far to an infected hive, which can take months to collapse. The reality is that the spread of this disease is largely our fault, with beekeepers using infected gear among different hives, moving infected hives throughout the landscape and letting infected hives collapse into a bee-deathtrap.

The frustrating thing is that the detection of this disease does not require any fancy equipment or complex techniques; it requires a stick. Typically, I will light my smoker with a matchstick and then tuck the extinguished match into my top pocket — this then becomes a highly calibrated investigation tool. If any capped brood cells look a bit off, I simply poke the used matchstick into it, and then draw it out. If a greasy, sticky rope of material comes out, then it is really bad news: the colony almost certainly has AFB. Anything else, and you most likely don't (but you might still have another disease).

That's it — a stick is all you need to shut down further spread of AFB.

I get suspicious if I see a colony that seems listless, where the foragers at the entrance seem to have lost their zing. But AFB can be present in hives that seem perfectly healthy. Unfortunately, they won't stay healthy for long if AFB has managed to gain a foothold.

If you have a suspected case of AFB, you need to get a lab test to confirm, but the symptoms are not subtle, and I tend not to waste any time waiting for results to come back — I will euthanize the hive and either burn or irradiate it to sterilize. We are lucky in Australia, as there are specific facilities where gamma irradiation is available to sterilize hives, so you get your gear back. In most other countries, burning the hive in a pit is the only way to deal with the infected material.

Euthanizing the hive is heartbreaking, but essential to stop the infection from spreading. You can either shake the bees into a large tub of soapy water (bees drown almost instantly) or pour a cup of petrol into the hive (only if you are burning). Either way, the dead bees and

The matchstick test for AFB is cheap and effective.

brood need to be buried, and the hive and any honey must be sterilized or burnt. One tip: the honey will expand during gamma sterilization, so read the instructions about filling and sealing tubs carefully!

So that is the bad news, and it is awful. But there is good news, too: the fact that this disease is mostly spread by us means that we are largely in control. Institute a barrier system so that no gear is swapped between apiaries. If possible, bring in brand new or sterilized gear. Always inspect your hives regularly, and especially if they seem listless. Simply put, there are no excuses — because you can always get a stick.

EUROPEAN FOULBROOD

European foulbrood, or EFB, is another bacterial disease of the brood, and is the other main disease that we are looking for when we inspect the brood chamber. It attacks the uncapped brood when they are in grub form, and before they pupate inside their cocoons. Thankfully it is less of a death sentence for the colony than AFB, but still represents a major problem that cannot be left unchecked.

Like its nastier American cousin, European foulbrood spreads via contaminated honey and gear, and the infection starts when a grub ingests spores into its gut, where the bacteria rapidly multiplies. The infection eventually takes over the gut of the larvae, starving it so that it curls upwards into a brown/yellow molten mass that stinks (hence the name).

This is another disease where regular inspections of the colony's brood chamber is essential, as well as good notes on previous inspections. The most obvious sign of EFB is uneven brood with patches of dead and dying baby bees. On closer inspection, you will most likely see the uncapped brood dead or dying, appearing molten or yellowing as they rot.

Sometimes this rot has a sour smell, but this won't be apparent in the early stages, which is when you want to catch this one. The larvae won't be curled neatly in the normal 'C' shape inside their cells, but will look sort of twisted or contorted.

You almost always have to do a laboratory test to differentiate this disease from AFB, as they can look really similar. Apparently the roping test has a shorter 'rope' length for EFB, but thankfully I don't have enough experience with either disease to make a call on that. I don't muck around — if I see something that looks off, I take a sample and get it properly lab tested. It is a relatively minor expense that can save your apiary, worth hundreds if not thousands of dollars. Not testing really is false economy.

The main difference between EFB and AFB are the potential consequences. Unlike AFB, a European foulbrood infection can be cleared with careful management that includes supplementary feeding, replacing the old combs and the diseased material on them with new combs, and improving warmth and ventilation for the hive. As with all the other diseases, preventing spread is paramount, and any removed combs should be burnt or irradiated. Robbing by other bees is a real potential with a weakened hive, so the colony entrance should be reduced and a robbing screen installed to prevent transfer of disease via the bees themselves. Even with these options, it is almost always a reportable disease to your local bee biosecurity officer, and they might still ask for the colony to be euthanized, especially where there are no other known diseased colonies and elimination is a possibility.

CHALKBROOD

Chalkbrood is a fungal disease of honeybees which weakens the hive and robs it of vigour, but generally won't kill a colony outright. Chalkbrood is picked up by foragers on flowers that have also been visited by bees from an infected hive. If the fungal spores reach the early brood they infect it, eventually desiccating it as a white 'mummy' that is covered in more spores. If the colony does not properly clean the hive, these spores then spread to other sections of the hive, infecting more brood and being picked up by foragers who leave the spores on flowers to restart the process.

This is one disease where often there can be visible signs of disease from outside the hive. As the dead brood starts to pile up, the undertaker bees will start to get overwhelmed. They will just toss the dead brood out the front of the hive, where they look like little white or dark pellets. This is often a sign of fairly advanced infection, but it is one area where rooftop

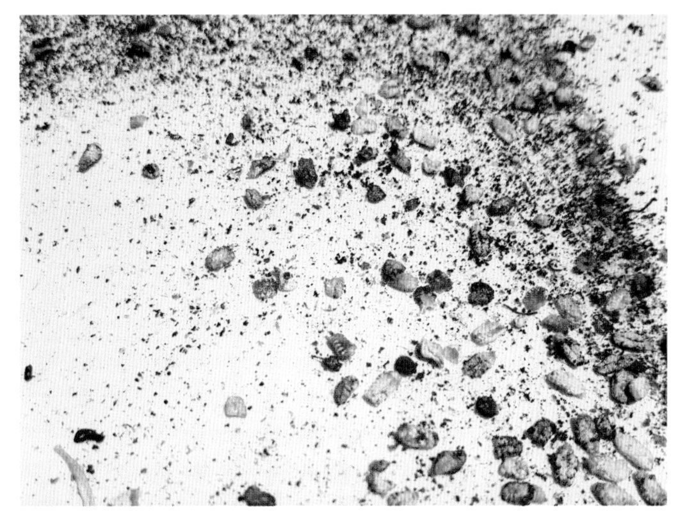

Chalkbrood 'mummies' on the base board, indicating a heavy infection.

White chalkbrood 'mummies' in among healthy brood.

hives have an advantage — the concrete and stone surfaces that hives are kept on make it a lot easier to spot the mummies, and birds seem less keen on eating them than regular dead bees.

The much better option is to spot the infection in the early stages, when there are only a few cells with chalkbrood mummies in them. They tend to stand out at this stage — the brood are clearly dead, and they are bright white and sometimes black against a backdrop of caramel-coloured, healthy capped brood.

With early infection it is often worth trying the 'banana trick'. Normally you should stay well away from banana when beekeeping, as it smells like one of the bees' alarm pheromones and puts the hive on an instant war footing — not exactly what you want when elbow deep in the colony! However, when treating chalkbrood, wait until the last minute, then squash half a banana on the top of the frames, close the lid and run! This stimulates an extreme cleaning response in the colony. They will chew up the banana and throw it out the front door, then clean down all internal surfaces, also throwing everything out the front door. If the infection is small and confined, this intense spring clean will usually resolve it. The front of the hive rarely has flowers around it, so this will not present any sites for the spores to hitchhike back inside the hive.

For serious infections, the cure is more drastic: replacing the queen with a hygienic strain. Hygienic strains of queen bees are raised by specialist queen breeders, and these queens create colonies with an obsessive cleaning instinct that keeps the interior of the hive pristine. All bees clean, but these ones really take it to the next level! Replacing the queen seems drastic but it will fix the problem, and will result in a much more productive hive than trying to struggle on with a weakened colony.

Some beekeepers seem to ignore a mild infection, but this is a big mistake, especially if you are in an area with diverse native bee populations. While chalkbrood weakens a honeybee hive, the disease is lethal to native bees, which have almost no natural resistance to it. Flowers visited by infected honeybees become a deathtrap for our natives, particularly Blue Banded bees.

SACBROOD

Sacbrood is a viral disease that is fortunately relatively rare to see as a major infection, but it can look superficially like some of the more serious brood diseases. It is caused by a virus that infects the brood through food containing the virus, which then causes the baby bee to die and dissolve into a fluid-filled sack. The key indicator for the virus is a conical point to

Sacbrood can build up in weak hives with poor hygiene.

the brood sac, and they are watery as they decompose, rather than being greasy or forming a sticky rope when probed with a stick.

The virus can build up in hives that are weak, particularly when they don't have good hygienic genetics. Changing the queen to a more hygienic strain is one solution, but beekeepers can also help out the colonies by cycling frames. Don't just leave frames in a hive year in year out; you should aim to replace them with fresh wax every three to four years. Some hive systems help you do this, particularly the Warre and top-bar hive systems, as the harvesting method also harvests the wax. As the bees are always building onto fresh wax, there is significantly less chance that disease will build up in the hive. Even for a Langstroth hive, you should aim to regularly take frames out of the brood box. This is most easily done by taking out the end frames once the bees have filled them with honey, moving centre frames out by one, and then giving them two fresh foundation frames in the centre of the brood box. Over the course of a season, you can replace the wax pretty easily this way; just remember that they are leaving the end frames full of honey for insulation, so don't undertake this process if there is likely to be cold weather.

SUPERSEDURE

This is not a disease, but it can create patchy brood patterns that look like many of the symptoms of brood disease. As the queen ages, she will start to run out of viable sperm — she only makes one mating flight, so once she starts to run out of the sperm stored in her body, that is basically it for her. As beekeepers, we can see this as an irregular brood pattern with random gaps throughout. The colony will notice this as well, and will start making arrangements to replace her, eventually either killing the old queen or having her leave with a last-ditch swarm.

The key difference between supersedure and disease is that there will not be any other signs of disease or decay in the brood. There will just be blank cells, with no corpse left behind. The gaps in the brood pattern will also be consistent throughout the brood chamber. Your record-keeping is also very handy here. If you are diligent with your inspections, you will see the gaps in the brood become progressively worse, until the issue suddenly resolves. This is the new queen starting to lay. Check carefully and you will probably find a shiny new queen bee, missing any paint marker the old one might have had!

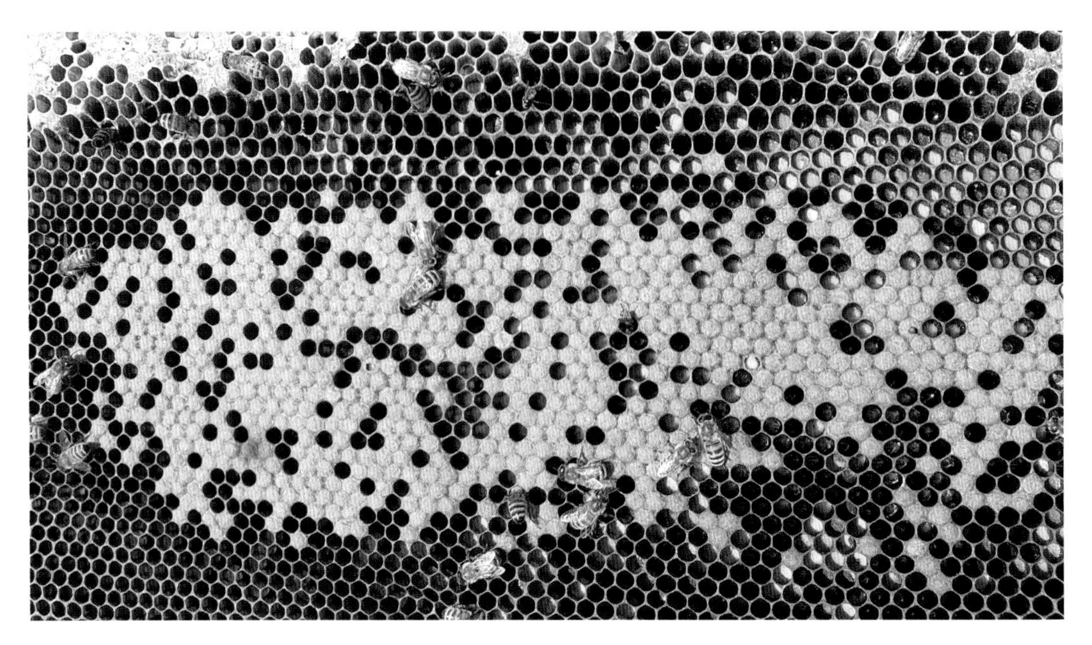

Patchy brood pattern, but no disease — the colony replaced the queen shortly after this.

Over time you will get a feel for when colonies are getting ready to replace their queen. Seeing more than the usual gaps in the brood frames is your signal to be careful on two fronts: check to confirm that it really is supersedure and not a disease; and also make sure that you don't accidentally squash or pull apart the delicate queen cells. Even if you plan to replace her later with a commercially bred queen, having a queenless hive is a headache that you don't need.

Sometimes something goes wrong and the colony is unable to raise a new queen. In nature this puts the colony into a death spiral, as they are unable to raise a new queen without fertilized eggs. This is where we earn our keep as beekeepers, bringing in a new queen to save the colony or giving them a frame containing eggs and young bees to let them raise their own. In a community garden, queenless hives can be crankier than usual — you want to get ahead of this problem by bringing in a new queen.

VARROA MITE

In late June 2022, a plan that had been years in the making swung into action. The objective was to achieve what no country had ever achieved before: to completely kill the Varroa mite within Australia's borders. Every other country had tried and failed to achieve this, but the payoff if successful would be well worth it.

When it comes to pests and diseases worldwide, the Varroa mite (*Varroa destructor*) is the really big one. It is a sort of bee tick that feeds off the bees, weakening them and spreading disease. Across most of the world this is the major factor weakening and even killing hives outright, and occupies much of a beekeeper's time. However, in Australia we have been blissfully keeping bees relatively effortlessly, secure in our status as the last major honey-producing nation free of this scourge.

This changed when the first detection of *Varroa destructor* happened near the port of Newcastle. There had been smaller incursions of the two Varroa species in the past, but this was the big one — the mites had been spreading for at least several months and had a decent foothold. The race was on to see if we could become the first country to successfully eradicate a major incursion of the worst bee pest in the world. Others had tried, with New Zealand coming close, but it has proven to be a fiendishly resilient pest.

The National Varroa Eradication Program was developed following the discovery of Varroa mites at the Port of Townsville in 2016, and was based around the one critical weakness of the mite: it must live with honeybees (known as an obligate parasite), and a honeybee cannot live alone; she must live with her sisters in a large, hungry colony that is relatively easy to detect. As a result, all of our early warning, monitoring and control is centred around bees and the beekeepers who manage them.

Varroa mite can only spread relatively slowly under the mites' own steam, as they hitch rides from bees

Bees being released back into the hive after a sugar shake test.

to hop from one colony to another, and sometimes on swarms. They can only move really big distances if we move them, and unfortunately moving thousands of hives from one end of the country to the other has become critical for much of our agriculture.

A ban on movement was immediately put in place, and over the next two months a small army of volunteers tested literally thousands of beehives to figure out exactly where the mite was. Once this was done, the heartbreaking task of killing all the infected colonies, as well as potentially affected ones in the surrounding areas, started. It says a lot about the character of the apiarists in these zones that they did not flinch when they were called on to make a horrible choice in order to save the wider industry.

What this means for Australia

With the initial containment appearing to be holding, the focus now shifts to all beekeepers to become more vigilant against potential infection. It literally takes years of painstaking testing before a pest can be declared fully eradicated, and it is critical that everyone, including urban beekeepers, becomes the eyes and ears of our biosecurity system.

Urban beekeepers have another key advantage in the fight against Varroa mite (and many other diseases) in that we are stationary and often a bit isolated from nearby beekeepers. This means we can get ready by establishing strong biosecurity barriers, and we have also bought ourselves time to start using the new breeding lines of Varroa-resistant honeybees. In many ways this most recent incursion has given us a taste of what might be coming, and the real work is just beginning.

Varroa management worldwide does not rely on a single approach. All of the most successful approaches use a range of methods, often within an integrated pest management framework.

Ongoing monitoring: A fact of life going forward

Whatever happens with the incursion in New South Wales, the one guaranteed feature of life for beekeepers going forward is ongoing monitoring. We will be testing for Varroa mite from now on throughout Australia, either for detection in new areas or for ongoing management if the worst happens and the mites establish.

From 2017 through to 2021 I was part of a team running the Sentinel Hive Program for the areas surrounding Canberra Airport, part of the national program that detected the Varroa mite incursion in Newcastle. It was a lot of extra work on top of my existing hives,

but I really believed in the program, and it has absolutely proven its worth with the current incursion. Early detection offers a great chance to eradicate the mite entirely. And even if eradication fails, early detection allows us to keep it bottled up while beekeepers are trained to manage for it.

To date, nobody has found the perfect 'system' that will prevent colony losses, and the mites always make beekeeping harder and more laborious, but it can be managed.

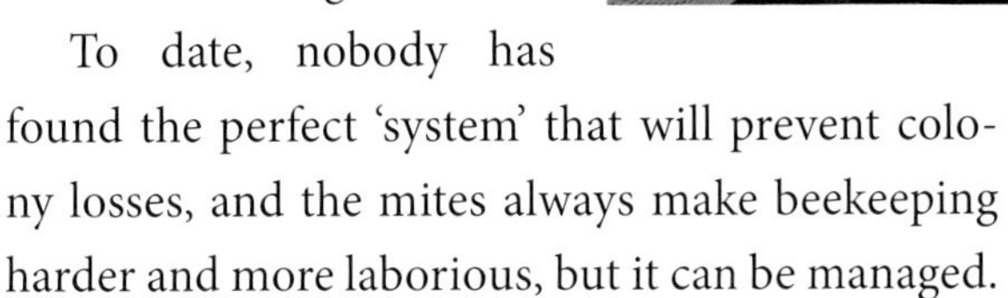

Monitoring methods

All disease control starts with consistent, effective, well-documented monitoring and Varroa mites are no exception. Even if you live in an area with no known incursions of mites you should be doing this for the practice, and in most areas of the country, it is mandated that you do a least a few 'sugar shake' tests per year; this is one of the four commonly applied tests to detect Varroa mite, all of which are detailed below.

Checking a mite board to detect Varroa in a Sentinel hive.

Sugar shake container for Varroa detection.

Sugar shake

This is just about the perfect detection test for broad-scale monitoring. It's high accuracy, non-lethal, and easy (and fun) to do. The test is exactly what it sounds like: you get a sample of bees and shake them around in some icing sugar. Once the

Bees walking back into the hive after a sugar shake test.

bees are coated in the fine granules of sugar, they become very slippery to the mites, which lose their footing and end up in the sugar. You sieve the bees out of the sugar, and then dissolve the sugar in water to reveal any mites present. The furious (but delicious) bees are dumped back into the hive to be cleaned by their sisters. Honestly, it is worth doing this test just to see the snowy white, angry bees become happy and content as they eat the sugar. I always get students on my courses to do this test, and we always end up having a chuckle. The main downside of this test is that it will miss any Varroa mites breeding inside capped brood, which is where most of them will hide early in the infection, but as a general detection test it remains the best in terms of reliability, safety and simplicity — all you need are bees, sugar and a sieve!

Alcohol wash

This is the less kind version of the sugar shake: washing a sample of bees in alcohol to kill the mites, which then drop off and are sieved out of the bees. Unfortunately, this also kills the bees, but it is favoured for biosecurity investigations as it has a significantly higher detection rate. As in the photo opposite, a sample of bees from the brood comb (where most of the mites are pupating) is collected into a jar and shaken vigorously for 4 minutes. I use a Tupperware container with a built-in sieve, and the alcohol kills the mites, which then get shaken off inside the container. The sample of bees is still relatively small (around 300 bees

Collecting bees into the alcohol wash container.

out of a colony of 10,000+) — just make sure that this doesn't include the queen!

This test also involves handling pure alcohol near a lit smoker, so there is the risk of fire as well. Always have a bucket of water handy just in case.

Drone uncapping

This exploits the fact that Varroa mites will specifically try to infect and breed inside the capped male bee (drone) cells. Rows of drone cells (easily distinguished by their raised conical caps) are pulled open using a special fork, and the pearly-white larvae are inspected for the darker mites. This also kills the male bees, but they are fairly expendable, and the colony can easily make more. Harsh, perhaps, but sacrificing a small number of drones to detect Varroa is well worth it, and this test will also pick up the Varroa hiding inside the capped cocoons of the drones.

Sticky mats

The only test that works without opening the hive, this is also by far the most unreliable and requires a special bottom to the hive. Mites on the adult bees will die naturally or be groomed off the adult bees by their sisters in the hive, and fall to the floor. Many hive designs now have a special screen in the bottom that allows pests like Varroa and Small Hive beetle to fall through, where they can be trapped on boards coated with glue or drowned in trays of oil. Regular inspection of these trays will reveal mites, but often only once an infection is well advanced. For the Sentinel Hive Program we would introduce miticides to the hives to increase the chance of mite drop, but this is not currently permitted in Australia.

Despite its limitations, this approach remains a useful passive detection method and is the only one that can be consistently applied through winter when it is too cold to access hives.

Management options: Mechanical methods

Once you have monitoring down pat, the next skill you need to learn is how to apply this to the actual management of Varroa if it ever becomes established. The most important management objectives for controlling Varroa are to ensure that all colonies have no mites heading into winter, and to recognize when to knock down numbers with treatments during the rest of the year. Having talked to a few beekeepers who are good at managing Varroa mites, it's evident that knowing how and when to create a brood break is an important skill.

Varroa mites must reproduce inside the capped brood cells, so the basic principle is straightforward: deny the mites any breeding habitat to reduce their numbers. In practice, some brood is essential for the colony, with literally thousands of new bees required every few days to replace losses. This is where the timing of the brood break comes in: it needs to be done during a nectar 'dearth', when little food is available. This happens periodically throughout the year depending on the mix of flowers available, and with experience, it can be predicted. During these times the queen will wind back brood production anyway so that there are fewer mouths to feed, so the impact of the brood break is reduced. (More on how to recognize and predict nectar flows later.)

The actual creation of the brood break is much more straightforward. It involves confining the queen so that she is still in the colony, but cannot get out to lay eggs in the comb. This is done using cages, meaning you have to find the queen during an inspection and then place her into the cage. These cages come in two styles, with a different strategy for each:

In-frame cages: These are smaller cages that fit in a cut-out section of honeycomb within a frame, allowing the colony to feed the queen but physically keeping her from laying eggs. The queen needs to be kept in the cage for at least two to three weeks, so that all capped brood have hatched, leaving nowhere for the mites to hide.

Whole-frame cages: These cover an entire frame, which allows the queen to lay eggs and form capped brood, but only on a single frame. All the mites will be forced to lay their eggs inside the single frame, which is removed on a weekly basis and the queen put back into prison in a fresh frame, which once again concentrates all of the breeding mites in a single frame, which can then be removed and frozen to kill them.

The end-result of both methods is the same: it forces all the remaining mites inside the hive to be outside of capped cells (known as their phoretic phase). This improves the effectiveness of Varroa-resistant bees that groom the mites off themselves, and in some cases this may be enough, especially when combined with a screened bottom board that traps the mites. Having all the mites out in the open, away from the protection of capped brood cells also makes any follow-up chemical controls much more effective.

Management options: Chemical control

Thankfully, I have never seen a Varroa mite in the flesh, and sincerely hope that I never do (at least in Australia), but I have applied a lot of Varroa control products! When I worked on the Sentinel Program, treatments were applied as part of the surveillance. Killing any mites that might have been in the hives meant they dropped onto the sticky boards we placed into these special hives.

I was really fortunate to have this practical experience under Australian conditions, and even before the latest incursion there was a lot of advice from overseas beekeepers on how and when to apply treatments. The use of chemical treatments is also one of the greatest sources of arguments among beekeepers overseas, and it often gets quite emotional. It makes it pretty hard to sort out what the real story is, but below I have done my best to summarize the most popular treatments based on my personal experience and from talking to experienced beekeepers. These are the treatments most likely to be approved for use in Australia if Varroa does escape containment and become widely established here.

The chemicals used to kill Varroa mite vary widely in their original source, how they are applied, and how effective they are. There are two broad categories, though: natural compounds and synthetic miticides. But don't let the 'natural chemical' tag fool you — cyanide and strychnine are also natural chemicals, and they will naturally send you to the morgue in no time. Always treat chemicals as a hazard to yourself and the bees, and *never skimp on the safety gear*! Most of these treatments will also contaminate honey, so your supers must be removed and kept off until the chemicals have left the hive.

Oxalic acid

This chemical is cheap, relatively easy to apply and naturally occurs in plants. This treatment cannot penetrate the brood caps, so it must always be used after a brood break (see p. 160) so that the mites are exposed as they cling to the adult bees. It is applied as a vapour (with a lot of safety gear to protect your eyes and lungs) or as a liquid that is drizzled directly onto the bees between frames. Bees seem to tolerate the treatment fairly well, and it appears to be pretty effective as an autumn/winter treatment.

Formic acid

Bees are fairly familiar with this chemical, as it is part of their venom! In fact, many insects (particularly ants) use formic acid as a weapon, often against other insects. In the Sentinel hives we used the Mite Away Quick Strips, which are a little like a strip of solid gel. The acid fumes build up in the hive and can go through the wax caps of the brood, killing mites at all stages. The main downside is that you must be very careful of the daytime temperature when you use them. If the days are too cool (less than 10 degrees C) the treatment doesn't properly infuse through the hive; if they're too hot (above 30 degrees C) it will kill the brood, and even the queen. We had a few hives lose their queens when using these, so it pays to have a reserve queen in a nucleus when using formic acid.

Amitraz

This is a synthetic miticide, which we used as infused plastic strips that were placed in between the frames in the brood boxes (two to three strips per box). They were easy to handle (you still need to wear disposable plastic gloves, *not* your beekeeping gloves) and had a long shelf life. The main issue is that amitraz permeates wax and any honey in the hive, and stays there for a long time. Its ease of use has made amitraz really popular overseas, meaning that the mites have had lots of practice developing resistance to it! When we were using amitraz strips in the Sentinel Hive Program, there were reports of high levels of resistance in New Zealand mites, so we would continually cycle the treatments to hopefully detect the mites somehow if they snuck in.

Other chemicals

I haven't covered the various essential oils, salts and acids that are also used, mainly because I have no frame of reference. These are also unlikely to be approved for use in the first instance; Australian authorities will go with what they know, and will start with the chemicals we have used in the Sentinel Program.

The use of chemical treatments should be a last resort; it is time consuming and requires special management and safety gear. All chemical treatments have pluses and minuses; however, they are better than just letting things rip if the mites do become widely established — we all saw how that went in us humans with Covid.

Reducing feral colonies

When a major event like the Newcastle Varroa incursion happens, it is natural to want to help, to take some action in your own area to assist in the fight. As it happens, there is one very effective approach that helps reduce the risk, not just from Varroa but from a whole host of diseases: eliminating feral colonies in the landscape. As the response to the Newcastle incursion progressed, there was an increasing focus on eradicating feral hives. With an estimated feral hive for every 1 to 3 square kilometres throughout Australia (and even more in some areas) this is a huge task, made harder for every swarm that is left unmanaged.[1] With no beekeeper to check them for disease, these feral hives can act as a reservoir for all sorts of disease, but one particular feature of the Varroa mite makes them particularly hazardous: the Varroa 'bomb' hive.

When Varroa mites become established in a honeybee colony, they will set to work parasitizing the brood and spreading disease over several years.

Feral swarm that took over a possum box. (Credit: Dermot Asls Sha'Non)

Eventually the weakened hive will start to collapse, with not enough workers to perform key functions — such as guarding the front door. Once this happens, other honeybees will start to notice and will be tempted to sneak in and steal some honey. Like the plot of a B-grade horror movie, the mites are lying in wait, now desperate to find a new host colony. They jump on to the robbers as they move through the hive, hitching a ride back to start their attack on a new colony.[2] While other diseases like AFB can work in a similar fashion, the thousands of mites inside a collapsing hive act like a bomb, spreading the infection to every colony unlucky enough to take the bait. As a result, every feral colony in our cities, parks and reserves is like an unexploded mine, just waiting for mites to enter to light the fuse.

Reducing, and even eliminating feral honeybee hives is very doable. With no beekeeper to care for them, they are vulnerable to queen loss, disease and hard seasons. We can lean into this to reduce the numbers further in a few ways:

Don't make more of them. Have the mindset that every swarm that escapes from your apiary is a threat. Do splits, regular inspections and always have a few swarm traps out during spring and summer — even if you don't have any swarming in your hives, these will catch stray swarms.

Remove and rehome swarms. Catch any swarms you hear about, and do cut-outs and trap-outs for established colonies (more on that later) for any established colonies in walls or in hollows. Don't just passively sit there to 'see what will happen'. We know what will happen: more swarms! Colonies can be successfully removed, even from tight spots like roofs and walls.

This colony had lived for several years in this roof space. (Credit: Dermot Asls Sha'Non)

Euthanize feral colonies. Tough, but if a homeowner, local council or park manager can't get a swarm out and want to kill it, support and encourage them to get it done. Honeybees aren't native, and if it's feral it's in peril in my book!

This is one of the classic 'no regrets' actions that we can take right now across the country, which will help us in lots of ways, and we can start today. Beyond denying Varroa a potential foothold we will be strengthening biosecurity against a host of other diseases, freeing up floral resources for production for your hives and leaving more room for biodiversity.

Hygienic and Varroa-resistant queens

Breeding queens with genetics that resist Varroa is the only long-term solution that we have, and selective breeding of queens has been the solution to controlling major diseases in the past. The devastating Isle of Wight epidemic, which was caused by tracheal (throat) mites in bees, started in 1904 and ran for 20 years, resulting in the widespread loss of honeybees in the United Kingdom. The response was a systematic breeding effort based on the surviving colonies, and this eventually resulted in the development of the famous Buckfast Abbey breed of bees. While these bees are not specifically bred to resist Varroa, Isle of Wight disease is now confined to the history books, and with any luck the Varroa mite will get to join it.

The obvious question is: why switch over to Varroa-resistant stock when we don't have Varroa? The answer is twofold. First, the selectively bred stock can be rolled out into our managed hives over years, instead of having to do it in a rush. The second reason comes from our most recent outbreak. Ironically, the incursion in Newcastle was close to the breeding station of Tocal College (an agricultural college run by the NSW Department of Primary Industries), where a trial was underway to breed Varroa-resistant bees as part of the Plan Bee Genetic Breeding Program. While a special shipment of carefully inspected bees was moved, to save them from the eradication efforts, it was too risky to start sending queens from the quarantine zones into uninfected areas. The time to move new queens out to the nation's beekeepers is *before* a biosecurity emergency, not after.

Plan Bee: Creating a systematic, science-based breeding program

Queen breeding has always been an exacting discipline within beekeeping, and creating Varroa-resistant strains of bees is now a national imperative. Queen bee breeders are specialist beekeepers who don't try to make honey or maintain thousands of hives for pollination; instead, they create queen bees and are continually trying to improve the quality of the genetic stock. This essential role was often carried out by separate businesses, and the Plan Bee Genetic Breeding Program was created as a national system for assessing the quality of queens. In addition, the program aims to bring an extra layer of science to the table with genetic and genomic testing. Without this, there is a risk that the drive to create the 'perfect' honeybee could see a reduction in overall diversity, making the national stocks more vulnerable to new pests and diseases.

One of these breeding lines was already specifically focused on resistance to the Varroa mite, with the behaviour known as uncapping investigated in the bees bred in the Wheen Bee Foundation research apiary on the Hawkesbury River, in New South Wales. Bees exhibiting this behaviour make a small hole in the caps of the brood, sniffing for Varroa and cleaning out infested cells. Tracing this to genetic markers will potentially allow this trait to be more easily integrated into the breeding programs at other apiaries. If a marker is found, this will be added to the genetic database that is an integral part of the Plan Bee program — and allow beekeepers like myself to be sure that we are getting genuinely resistant bees in the absence of Varroa mite in Australia. With the recent outbreak the focus on Varroa resistance will increase, so that if we do get another incursion, or if this one escapes, we are prepared with resistant stock.

Over time the database of genetic merit will help queen breeders and beekeepers identify the best stock, while also checking that we are maintaining genetic diversity. This is going to be important, as the price for these more highly bred queens is likely to be higher. People will want to see what they are getting for the extra dollars.

Another important part of the program is outreach to both beekeepers and the pollination industries, continually checking to see what traits are most important in honeybees.

There are several mechanisms for how bees attack Varroa mites in the hive. Some of these are compatible with productive beekeeping, some are most definitely not! One of the unwanted behaviours is how their original host, the Asian Honeybee (*Apis cerana*), deals with them. They naturally have open holes in the caps of their brood cells, and when they smell too many mites infecting the brood they will grab the honey and do a runner as a whole-colony swarm — known as absconding. Almost any disturbance will trigger this mass flight, including inspections by beekeepers, meaning that they are basically useless to keep in managed hives. Some subspecies of honeybees, particularly sub-Saharan and Africanized honeybees, seem to be able to tolerate Varroa fairly well, but they do not exist in Australia, and that is a good thing as they are extremely defensive.[3] Many people will know Africanized bees by their nickname of 'killer bees', and they are definitely not compatible with community gardens!

More useful traits are grooming and uncapping, which are the focus of selective breeding efforts worldwide. Grooming is where the adult bees attack the mites on their sisters, tearing them off and throwing them onto the floor of the hive.[4] Sometimes the mites will be damaged enough that they can't climb back on, but beekeepers can also help out their bees by putting a screened bottom on their hives. This lets the mites fall through into trays of oil or onto sticky paper, trapping and killing them.

Uncapping (also known as Varroa-sensitive hygiene) is a specialized behaviour where the nurse bees looking after the brood make small holes in the caps of the brood cocoons to sniff out signs of disease or mites. Diseased or infested cells are cleaned out, killing the mites before they can breed.

Finally arriving at truly resistant genetic stocks of bees will be a long-term journey over decades, and will span continents. Literally thousands of scientists, beekeepers and farmers are working towards this final fix. They are going to need our patience and support, but we have done this before with other pests and diseases. Eventually, this will happen for Varroa as well.

SMALL HIVE BEETLE

I have to admit to a grudging admiration for the Small Hive beetle. It has to be one of the sneakiest, most underhanded and courageous critters around. They have a uniquely clever strategy for attacking a bee colony, and without intervention from beekeepers in warmer

Small Hive beetles inside a weakened hive.

climates, they will often be successful in destroying the entire hive. Originally from Africa, they are believed to have snuck in on cargo shipments to Richmond Air Base near Sydney and have quickly established throughout the east coast of Australia.

The infestation of a beehive by Small Hive beetle (SHB for short) starts with a single beetle flying up to 10 kilometres once they smell out a hive, and getting past the guards at the entrance. The guard bees will try to stop them but they are incredibly tough, encased in sting-proof armour plates. Their chief strategy is to run past the guards and find a crevice or gap that the guard bees can't get into. They then hunker down and wait until the bees lose interest. Sometimes the bees will try to entomb the beetles in a 'jail' using propolis, but the beetles have evolved an ingenious way around this, too. They mimic the taps that baby bees make on the antennae of nurse bees, which tricks the guards into feeding them.[5]

Once the bees stop paying attention, the real damage is done — the beetle lays its eggs through the honeycomb, which hatch into maggots that eat through the honey and pollen stores. As they eat, the excrement of the maggots contains a yeast that makes the honey ferment into a pungent slime. The sickly sweet smell of this is a tell-tale sign of heavy infestation, and will draw beetles from near and far, who swarm into the hive to lay even more eggs. This creates what is known as a 'slime out' where the maggots turn almost all the honey stores into a fermented mess, ruining the interior of the hive. Eventually the bees will stop trying to clean the mess and abandon the hive to the beetles. Thousands of maggots then pour

Base trap of a Flow Hive 2 loaded with diatomaceous earth.

out of the hive, pupate in the soil and then hatch into adult beetles, which fly off to attack further hives.

The key to stopping this attack and saving your hive lies in that initial hiding attempt by the beetle. They are continually harassed by the bees while inside the hive, and will seek refuge in a crevice or gap. The trick is to lay a trap that looks like the perfect hiding place.

There is a wide range of traps that beekeepers use to kill hive beetles, and they all rely on working with the bees to herd these nasty critters to their doom, which I totally love. Here are the main types:

Base board traps

These have a slotted base that the bees can walk over but which beetles will fall into, with a reservoir that is filled with cooking oil (to drown them) or diatomaceous earth (to desiccate them). These are usually built into the base of the hive and have the added advantage of being able to be opened up and the trap tray removed in hot weather for extra ventilation. Some hive designs (such as the Flow Hive 2) have these as standard, but you can also buy them for Langstroth and Warre hives.

Beetle harbourage traps

These are cassette-style traps with a slot in them that looks like the perfect hiding place for the beetles as they scuttle into the hive. However, there is a nasty surprise in the form of either diatomaceous earth or cardboard impregnated with toxin. Placed in the bottom of

Beetle harbourage trap — an excellent control option.

the hive near the entrance, harbourage traps are an extremely effective way to control these pests.

Chux cloth

This is a folded piece of cloth that the bees will tease out into a fluffy mess. Placed in the top of the hive, it looks like a great option for the beetles, but when they try to hide in the folds of the cloth, the spurs on their legs get caught and they can't free themselves. A super-cheap approach, but this also sometimes entangles bees, which is not so great.

Between-frame traps

These are narrow containers which slot between the frames at the top of a Langstroth hive, and have holes in the top that allow the bees to chase the beetles in. Filled with cooking oil, the traps then drown the beetles and you can easily see how full they are during inspections.

Entrance traps

These are attached to the hive entrance and have a slotted screen that the bees can easily walk over, but that beetles will fall through, landing in a kill chamber filled with oil or diatomaceous earth. Highly effective, entrance traps help the guard bees do their job of keeping the beetles out entirely.

For rooftop beekeepers, there is another highly effective strategy that has an almost 100 per cent success rate: simply place your hives on a section of roof that has only concrete or coarse gravel substrate with no soil. The beetles must have soil to pupate, and while this won't stop them from sneaking in from other areas, it will prevent them breeding within the apiary itself.

One thing to note: while this exotic pest is a real hassle for beekeepers, it hasn't been all bad. The high prevalence of the beetles in wetter coastal areas has seen the hive beetles act as an effective biocontrol agent. They will try to attack native bee colonies, but they

Beetles need soil to breed. No soil = no more beetles.

are generally highly specific to honeybees, which are their main host in their natural environment. The result has been that well-managed honeybee colonies are protected by their keepers, while feral honeybee colonies are removed from the environment. Not a great way to get this done, but on balance a win for the environment, with more hollow trees freed up for native species.

BRAULA FLY

More of a weird, creepy annoyance than a serious threat, Braula fly (*Braula cocea*) is a wingless bug that looks superficially like a tick or a louse on the adult bees. The adults cling to the bees and steal food from their mouths as they feed, and then cling to them as they leave the hive, spreading to other colonies on swarms and due to bees 'drifting' into nearby colonies.

The problems to beekeepers are confined to Braula fly larvae, which tunnel through the honeycomb, creating a distinctive 'cracked window' effect that spoils the appearance of cut honey. While the larvae themselves won't spoil honey like Small Hive beetle, it still looks off-putting.

Treatment is surprisingly simple. The adults are extremely susceptible to tobacco smoke, which poisons them. As little as a few grams of tobacco in the smoker will produce enough smoke to kill the adults, and any frames with evidence of larvae can be frozen for 48 hours before being extracted to kill the larvae.[6]

EUROPEAN WASPS

If bees are the easygoing, hippie vegans of the insect world, then European wasps (*Vespula germanica*) are the street punks. An introduced pest from Europe, they have spread to all parts of Australia and many other parts of the world. When bees diverged from wasps around 70 million years ago, they left behind their carnivorous diet, but European wasps (or Yellowjackets in the US) definitely didn't! At certain times of the year they will relentlessly harass your beehives, trying to carry off brood or even adult bees. The wasps are after meat, specifically the succulent baby meat of your brood to feed their baby wasps. This constant harassment can turn hives excessively defensive — just imagine how you would feel about visitors if there was a constant stream of burglars assaulting your house!

This relentless hunger for meat is their great weakness, though; they are really easy to trap, and best of all the traps won't harm your bees if baited and placed correctly. Traps are generally effective at catching them, and I place them at 10-metre intervals around my bee yards.

European wasps are a hassle for your hives that you can easily deal with.

Wasp trap bait

Preparing wasp traps at the community garden apiary.

1 small tin of cat food (tuna works really well)
2 tablespoons fruit jam
1 tablespoon caster sugar
1 teaspoon baker's yeast
1 squirt of dishwashing detergent

Mix the ingredients in a sacrificial container (it is awful) then pour into the wasp traps. I find that the best ones are purpose-built units from gardening stores, but there are lots of designs for homemade versions using bottles, jars and other containers. The wasps are drawn to the smell of fermenting meat and jam, and the detergent breaks the surface tension of the mixture, making them sink and drown more quickly. The fermented meat smell repels bees, but it is worth checking to make sure that you are only catching the exotic European wasps, and not any native ones.

The most critical time to trap wasps is autumn. This is when they are trying to raise queens who go out to found new colonies. The more wasp foragers you can kill, the fewer queens they can raise. In late autumn and early spring you might even get lucky and catch the queen herself — an instant removal of a potential colony.

If you are lucky enough to spot their nest, call a pest controller to remove it. Be wary of trying to do this yourself. Wasps are extremely aggressive to anyone near their nest, and are much better armed than bees, able to sting multiple times. This is one job best left to the professionals.

Wasp traps protecting the hives.

Ant tunnelling damage to the walls of a hive.

ANTS

Ants love sweet things, and the honey from your bees is definitely no exception! In addition to the sweetness of your hives, the warmth and scraps of pollen that the bees drop around their hive attracts ants. As with most nuisance pests, ants aren't such a big deal for a strong hive, but they do annoy the bees and can make them cranky.

Warre hive keepers in particular need to watch out — ants love to set up shop in the quilt box, where it is humid and warm. They will react enthusiastically to any disturbance, and I had one Warre hive where they would run all over my hands as soon as I tried to inspect, biting me when they got into my gloves.

The good news is that they are easy to control. Ants hate cinnamon; it is a powerful desiccant (drying agent) and can kill them in large quantities. The bees don't care about it, so just sprinkle it around the hive every few days until they get the message. For particularly persistent ant problems, place the hive on bricks sitting in barbecue trays filled with water; even the most intrepid ants can't swim.

For your wider garden, excess ants can indicate there is an imbalance in your local environment, so try spreading mulch and placing bark and logs to create habitat for lizards, frogs and other predators to keep them in check.

PESTICIDE POISONING

Having a hive injured or killed through pesticide poisoning is an infuriating and heartbreaking experience. Pesticides can be an important tool in certain situations, but their overuse has become a serious problem in both cities and rural areas, although often for different reasons.

In rural landscapes, most of the people handling chemicals are trained to do so and any impact on hives is often the result of an ignorance of hives in the area or a miscommunication. In urban areas, most pesticides are being handled and used by people with little or no training. Gardeners might have heard that some chemicals can impact bees, but often don't know how to correctly prepare dosages, how to apply chemicals to reduce risk, or think that 'a bit extra' will work better. The result is a deathtrap for bees in someone's garden, with perfect-looking flowers hiding a deadly cargo in the nectar that the bees bring back to the hive.

The main symptom to look out for is a large pile of dead bees at the entrance of the hives, sometimes with dying bees doing a drunken, staggering walk. This is rarely subtle — the pile of bees almost blocks the entrance and sometimes needs to be swept up and disposed of (double-bagged to landfill) by the beekeeper.

A poisoned hive can recover, but it really depends on how much poison was carried back to the hive and whether the queen survived. You need to watch carefully for robbing, and reduce the entrance size right down. Robbing by other bees will not only impact the poisoned hives but could poison other colonies. Nursing them back to health will take time, but is doable with patience and care.

BUSHFIRE SMOKE

While not your typical pest or disease, bushfire smoke is a symptom of a sick planet and what we have done to our climate system. During the latter half of 2019 and early 2020, Australia experienced what became known as the Black Summer. An unprecedented weather system caused the most widespread and intense fires in recorded history, blanketing

A tell-tale sign of pesticide poisoning is a large pile of bees at the entrance.

Bushfire smoke isn't just dangerous for us.

vast areas with toxic bushfire smoke that lasted for months on end with no respite. In my hometown, New Year's Day to the end of February was especially intense, and Canberra had the dubious honour of having the worst air quality in the world for most of January.

It was awful to be stuck inside your home next to the air purifier, and having to wear a gas mask to go outside, but the bees had no such technology. Some researcher friends called to warn me that our bees were most likely being killed in large numbers. They had spent ages painstakingly marking 800 foragers for a study on their hives. The foragers then went to work, but only a scant handful returned. Losing over 90 per cent of the foragers is bad for one day, but sustained for months this would almost certainly be deadly. I immediately started feeding the bees sugar water and pollen substitutes to try to offset the losses (and keep them in the hives) and was able to save my colonies, but they took a huge hit and it was a close thing.

As I write this, similar smoke from Canadian fires is blowing over large swathes of North America, blanketing entire cities with choking smoke that is horribly familiar to many of us in Australia. It looks as if this is a hazard that we need to prepare for — but it also highlights the risk that climate change brings to our lives.

ENVIRONMENTAL IMPACT OF BEEKEEPING

When I started beekeeping it was a very 'niche' hobby, mostly pursued by bearded men and tough-as-nails old ladies. I think there were about 30 people at the first beekeeping club meeting I went to. A few years later I was president of the same club (followed by my friend Dermot, whose photos you can see in this book), we had just topped 500 members, and our meetings had to move into a larger hall to accommodate the 200+ people attending. Every

Beekeeping has become an increasingly popular hobby.

day another celebrity would take up the hobby, and our training team were run off their feet with courses, field days and demonstrations.

However, with this explosion in beekeepers came a realization that there are very real impacts from saturating the landscape with an introduced species. I still feel strongly that, on balance, beekeeping has been one of the most powerful ways to raise awareness about sustainable cities, pesticide risks and climate change, but the potential for negative impacts is real.

This leads to an awful lot of snarky takes on social media, especially by native bee enthusiasts, about how 'we are saving the wrong bees' and how honeybees are not under threat (more later on why this can be a self-defeating take). Rather than get defensive, it is a lot better to think through the issues being raised and reflect on what you can do to make sure they don't apply to you. The first essential step is realizing that the critics do have a point: there is almost certainly an impact on native species, and feral swarms are a form of pollution to our native areas.

The good news is that for each of the environmental issues, there are very real, practical steps that all honeybee keepers can take to proactively manage these problems before they arise.

BUZZ OFF: COMPETITION FOR FLOWERS

Commercial beekeeping involves moving thousands of hives from one end of the country to another chasing one of two things: nectar and pollen. It represents literally millions of honeybees on each truck suddenly appearing in a patch of flowers with no warning for the natives. It's pretty obvious that this will overwhelm the locals! However, some urban beekeeping is not far behind, with an increasing and sometimes unregulated increase in hives. When I started keeping honeybees a little over a decade ago, there were less than 100 members of the local bee club. Just a few years later, we had more than 500, and many of these people had more than one hive (some many more). This trend holds true in lots of cities all over the world — beekeeping with honeybees is a fun, fascinating, productive hobby that improves your garden.

This degree of increase in a supergeneralist pollinator, able to visit all sorts of different flowers, was always going to have some sort of impact on native bee populations. The interesting thing is that it is not uniform. There is clear competition for flowers between native bees and honeybees when they are around the same size, and less evidence of competition with the smaller native bees.[7] Native bees have also evolved with native plants (obviously) and seem

My rooftop apiaries have sometimes hosted significant numbers of hives.

Native species have beautiful flowers, but an important function to go with them.

to be better able to exploit these resources.[8] The good news is that, at least in some Australian studies, the native bee richness seems to be holding up within urban environments, with honeybees not contributing to an obvious decline in native bees, despite the increasing popularity of beekeeping.

It is fairly easy to put limits on the number of hives in cities. Almost everyone who keeps bees in an urban area does it for fun, so they don't need to have all that many hives — nobody is going to go broke if they can't keep hundreds of hives. In my local area, the number of hives you are allowed to have is linked to the size of your property, but maxes out at a certain number (usually around eight), so there will never be the thousands of beehives concentrated in a small area that you would see in some agricultural areas. At the same time, it is worth noting that these concentrations of hives in agriculture only happen for short periods, whereas backyard hives are always present in the landscape. Keeping a relatively low density of hives means there are always enough flowers to go around — good for all bee species, and good for beekeepers.

Recently the sustainable carrying capacity for honeybee hives was figured out, and it seems to be around 7.5 hives per square kilometre of urban greenspace — obviously more greenspace with flowering plants supports more bees.[9] A lot of legal controls on honeybee keeping limit the number of hives to a certain number per block, but there are no limits on how many homeowners can be beekeepers. Ideally, we should be keeping tabs on the density of hives across the landscape, as this is the metric that actually matters.

The second big change is the increasing willingness of local governments, community

Swarms are great for starting new colonies, so long as they are in a managed hive!

groups and the wider public to plant pollinator-friendly flowers. More than just a sudden rush of nectar in spring, increasingly the flowers being planted bloom throughout the year. Beautiful and bountiful resources create something people value and want to retain, so this is something that is going to persist for years, hopefully decades. It is also an easy message to spread: just plant flowers. And then plant some more, ideally ones that will bloom when other flowers aren't out. Then start planting more native plants, in all different varieties and flower shapes. The larger and more diverse the plantings of native flowers, the better for our native bees. This will still provide nectar for honey, but this diversity in native bee populations (not to mention moths, butterflies, beetles, ants …) is what we should be aiming for. Native bee researcher Dr Kit Prendergast has studied flower preferences extensively, and her research shows that not only are native plants preferred by native and honeybees alike, but that a higher density of native plants allows native bees to shift their foraging to cover a wider range of plants.[10] Having denser plantings of native flowers also boosted the breeding success of native bee hotels.[11]

Don't just let it go: Uncontrolled swarming!

Hollow-bearing trees are one of the most critically low resources for native species in many environments around the world. Every one of these that is occupied by exotic honeybees is a potential breeding site lost to native species. Even though at the moment we don't really know what sort of impact this is actually having,[12] it is definitely not good.

While it is fairly obvious that a tree hollow that has been occupied by honeybees is not available for birds and animals (even really tough ones), the actual impact of bees in the wider landscape is still a bit of an unknown. I have personally observed wild bee colonies occupying hollows in trees that were previously recorded to have bird nests, but there is little systematic evidence that the native species were not able to find alternative hollows. Tree hollows are also sometimes taken over by native wasps, which also aren't the sharing types, so this type of invasion is not unknown to native species, and is it reasonable to expect that they have evolved to adapt to this event.

What is certain is that we have precious little native vegetation left, and in many areas widespread land clearing has made large native trees that can form hollows a relatively rare commodity. In this situation any loss of a hollow will have an impact, and in these areas your responsibility as a beekeeper will be greater. Like most environmental management challenges, the actual impact and the viability of mitigation is highly dependent on the situation, and there is a lot of good that you can do. As a nice bonus, it is fun to catch swarms and keep them from becoming a problem.

One of the actions that scientists and governments are taking in many areas is the installation of artificial nest boxes. Sometimes these have low occupancy rates with native species, who prefer the real thing, but they do have some big advantages. In the case of

A very unhappy possum being forced out of its home by a swarm of bees. (Credit: Dermot Asls Sha'non)

honeybees, there is a lot that can be done to prevent buzzy interlopers from moving in. Treatments such as charring the upper lid of the box and scent deterrents have proven to be highly effective in keeping bees out. Bees generally won't attach their comb to charred surfaces, and the scents we use for bee removals persist in wood for a long time.

I have a range of nest boxes in my garden, from a box designed for Pardalotes (a beautiful native bird that I love) to bat and small bird boxes. If you have room, it is worth putting up as many as you can, ideally in areas where domestic cats and other predators cannot reach. Just remember to char the inside of the lid, so that it stays reserved for native species.

Going viral in a bad way: Disease transfer

Chalkbrood is a common fungal disease of many honeybee colonies, but it is not a particularly serious disease (see p. 150). It will weaken a colony by infecting the brood with fungal spores which desiccate them, creating the characteristic chalky white 'mummies' that are the key diagnostic. These mummies are often left at the entrance and are loaded with fungal spores, which foragers then pick up on their feet and transfer to flowers and watering sites, where other bees come into contact with them and are infected.

Honeybee colonies are generally able to push past a chalkbrood infection, with the loss of even a few hundred larvae not fatal to the colony when the queen can lay upwards of a thousand eggs per day. However, for native bees, which typically lay less than 30 larvae in their nest, this is a disaster. Chalkbrood arrived in Australia on exotic honeybees, and native bees have almost no immunity to the disease. As a result, carelessly allowing chalkbrood to persist in your hives can condemn native bees in a wide area to death when they pick up chalkbrood spores on flowers and watering sites visited by honeybees from infected colonies.

Chalkbrood is fairly straightforward to manage by maintaining a good barrier system to prevent movement of infected gear, and avoiding cold, damp sites. Honeybees will clean out infected cells, and hygienic strains of queen bees are available that will stimulate more intense cleaning behaviour in the colonies. As a last resort, a strip of banana peel left inside the hive at the end of an inspection will drive the colony crazy, and they will intensively clean the interior of their hive to remove the smell, and in the process clean out any chalkbrood spores.

10

Swarm catching for fun and profit

Different methods to bag your bees

I remember the first time that I experienced a swarm of bees. I was about thirteen, riding my bike over a wooden footbridge on the shore of a lake. Suddenly there was a roaring sound, and a swirling mass appeared at the other end of the bridge. Apparently, the bees didn't feel like crossing over open water and, like me, had also decided that the bridge was a great way to get from A to B. I had no idea what to do, so just hit the deck as the bees roared over the top of me. I was more fascinated than scared, but I still expected to get stung. As quickly as they had appeared they were gone. The whole experience took less than half a minute and I was left with not a scratch, just the shock of having had a force of nature pass over me.

The swirling, roaring mass as tens of thousands of bees take wing always generates fascination and media attention every swarming season, which generally runs from spring through to mid-summer. For beekeepers it can be a headache as this is half your workforce heading off into the wild blue yonder, potentially to bother the neighbours. As noted previously, there is a serious side to this as well: swarms settling outside of a managed hive are an

Parliament beekeeper Sarah Asls Sha'non catching a swarm in the garden.

Honeybee swarms are loud and spectacular when in flight.

Shaking all the adult bees out of the hive to create a 'shaken swarm' will sometimes work.

environmental and practical problem, especially in urban areas.

Swarming is both a natural part of honeybee biology and a major cause of headaches and controversy for urban beekeepers. Becoming good at preventing swarms through regular inspections and splits is a foundational skill for honeybee keepers, but expecting to be 100 per cent swarm-free is like expecting teenagers to not have premarital sex. Luckily the results of swarming are easy to fix, and won't hang around for eighteen years refusing to clean their room.

SWARM BEHAVIOUR

A honeybee is by nature a communal creature. A single bee separated from the hive cannot survive for more than a few days on its own. Everything they do, they must do as a collective, and establishing a new hive is no different.

We still don't understand what triggers the instinct to swarm, but once the colony has made this decision, there is precious little that a beekeeper can do about it, so you need go to with the flow, and ideally split the hive or do a 'shaken swarm' to let the colony get the urge out of its system. However, when this doesn't work, half the hive will take to the air in a spectacular, loud swarm to establish a new colony.

I love swarms. They have a real beauty and drama to them, from the swirling, roaring chaos of the bees in flight to the smooth cluster of bees as they protect each other from the elements. People often react with fear, but the reality is that most swarms are incredibly gentle. There is a special feeling to being around a swarm when it is on the move. It is a maelstrom of pure, vibrant life as the bees strike out to find a new colony.

While as beekeepers we try to prevent swarming, it doesn't always work, so being able to catch a swarm and put it into a managed colony is an essential part of your skillset. The best thing is that everyone watching will think you are some sort of weird, crazy magician, when in reality it is one of the easiest things to do in beekeeping.

I started catching swarms fairly soon after starting beekeeping — there was a shortage of swarm catchers, and one of my mentors was an absolute magician at it. He can basically do it all, from the basic shaking bees off a shrub into a box, to being suspended on a cherry-picker cutting them out of a building with the help of thermal cameras, power tools and a bee vacuum (yes, that is a real thing). I soon found myself running around the suburbs collecting swarms from gardens, compost bins, walls, even public playgrounds and cars. However, my all-time favourite place to collect swarms from is trampolines. For some reason they love to cluster on the underside against the dark fabric. It's frivolous, whimsical fun — all you have to do is set a hive up underneath them, and then release your inner child, jumping on the trampoline to shake them down into their new home. Luckily the homeowners usually can't see my big goofy grin behind the bee suit veil as I do this.

So yeah, it is kinda fun and fairly easy, but there are a few tricks to know that make collecting swarms easier on the bees and the beekeeper.

HOW SWARMING WORKS

It is well worth understanding how swarming works, both for your hive management and for when you have to catch a swarm. A honeybee is by its nature a social creature. Unlike their solitary native bee cousins, the queen cannot just strike off on her own to found a new colony; she simply must have her entourage! Like almost everything else that bees do, it is a group event that is very well planned.

The instinct to swarm starts several weeks before (possibly even longer) in response to a range of cues that we don't fully understand. The key triggers seem to be a combination of the following:

Nectar flows, with a large, intense nectar flow commonly stimulating honeybee colonies to swarm. This makes sense when you consider that the swarm has to set up shop from scratch, so abundant resources will be essential. Some Eucalypt species are known to create intense nectar flows that commonly stimulate swarming.

Lack of space, particularly when honey storage starts to encroach into the brood nest where the queen is trying to lay eggs, which is known as the colony becoming 'honeybound'.

Swarms often settle in trees while they figure out where to go next.

The departing swarm takes some of this honey with them, creating space.

Older queens, as swarming is dangerous in nature, with a high chance of failure. As a result, older queens that are starting to run out of fertilized eggs are more likely to take the chance, as they are likely to fail soon anyway. This is one reason why some beekeepers replace their queen every year, but it is not foolproof: if there is a lack of space and/or lots of food in the environment, the colony will still swarm.

Once the colony has decided to swarm, the bees prepare special chambers along the bottom edges of the honeycomb where an egg is placed surrounded by royal jelly. This is a

A swarm from one of the community garden hives.

special food that transforms a normal worker bee larvae into a queen bee. They will often make lots — their queen is about to leave, and they don't want to be left without a new queen. Once the baby princesses are formed and close to hatching, they will start making a 'piping' noise inside their cell. This is the signal to the colony that they are healthy, and this is likely to be one of the 'go' signals for the swarm to form up and leave the hive.

As the swarm forms, the colony splits down the middle, taking an even slice of foragers, guards, young bees and scouts. All the workers 'pack their bags' by eating as much honey as they can before heading off into the wild blue yonder to find a new home. The old queen also leaves with the swarm, leaving behind developing princesses to carry on the colony. The roar as the swarm leaves is unmistakable and is why people are often so afraid of them; it is a primal sound that is unlike anything else, almost like a jet engine, but made up of a swirling mass of bees. If you haven't been doing inspections, this is likely the first sign that the colony is swarming. Not ideal.

The swarm will leave the hive and cluster together, hanging out while scouts leave to find a suitable home. The scouts will check potential hollows and are pretty exacting, measuring up any space they find to check that it has enough room (they actually pace it out to calculate the volume; it is an amazing skill). They then return to the cluster and dance to give the direction and distance to the candidate hollow. The swarm sends out multiple scouts to assess as many options as they can find, and their enthusiasm in dancing depends on how good a fit the space is. Once enough dancers have agreed on a good prospect, the swarm makes a decision and moves in. This is a really important facet of their biology — the better a fit your swarm box or bait hive is to an 'ideal' home, the more likely it is the bees will do most of the work for you.

People tend to notice the bees when they are in flight or when they have clustered somewhere. This is usually when beekeepers get the call that there is a cluster of bees, and it is then a race against time to get to them while they are still deciding where to go. If a beekeeper can reach them in time, we can short-cut their deliberation of which home to move into and give them an even better option: a hive box that is perfectly set up, ready for them to move in! It doesn't always work that way, but that is the ideal and happens most of the time when we can reach them while in a cluster.

The second key part of swarm behaviour is that they want to stay together, and ideally clustered around the queen. This allows us to get the entire swarm into a single box, as once you have the majority of the bees in, the rest will want to follow to stay with the group. This includes the queen; while they cluster around her she will try to stay with the main group as much as she can, so she will generally be inside the largest mass of bees in the cluster. Once the queen and most of the colony is inside, the bees at the entrance release a 'homing beacon' scent (Nasonov pheromone) to let the rest of the swarm know that 'we are all in here'.

Bee fanning Nasonov pheromone into the air as a homing beacon for the swarm. (Credit: Dermot Asls Sha'non)

Once they are in the box, I like to shut them in (with lots of ventilation) for a day so that they 'lock' onto their new home and start raising brood. There will be an explosive building of new honeycomb so that the queen can start laying the next generation of bees. I often place developing swarms in orchards that I am pollinating. They are so driven to set up their new home that they work incredibly hard, and fruit blossoms have all the nutrients they need. The organic orchard I work in has a variety of fruits and nut trees growing, with understorey herbs left in place, giving the bees a rich variety of flowers to forage in.

The next part is why many experienced beekeepers prefer to establish colonies with swarms. The swarm has all the workers, cleaners, nursemaids and foragers and has been producing wax in readiness for creating honeycomb. The genetics of wild swarms will be from strong local colonies that have adapted to the local environment, and often form robust, productive hives. Wild swarms often take over important tree hollows, so housing, collecting and nurturing these swarms that would otherwise be an exotic pest is one of the ways beekeepers can limit the impact of honeybees on the environment.

SWARM TRAPS

The easiest way to catch a swarm is to get the bees to do all the work for you, so you hardly have to lift a finger. Swarm traps are nucleus or even full-sized hives that are set up to catch swarms. The term trap is a misnomer — it isn't a trick, but rather a hive ready for them to move in, with the front door open and nice furniture already set up. While these come in all shapes, sizes and configurations, there are a couple of common elements that make an effective swarm trap.

Using a nucleus as a swarm trap

I have a nucleus set up near my apiary as a swarm trap from spring until the end of summer, and every single year so far I have received a gift of free bees in the form of a swarm. While swarm management is a critical part of good bee husbandry, sometimes what we want and what the bees actually do are two different things. As a result, having a Plan B swarm trap set up can be the difference between catching and losing a swarm.

A swarm trap is basically anything that attracts swarms of bees, making it easier to move the swarm into a managed hive. The bees can freely come and go; it is less of a trap and more clever real-estate salesmanship on the part of the beekeeper. As discussed earlier, swarms

Success! A swarm moving into a bait hive.

Nucleus hives for swarm-catching don't have to be fancy — I made this one from scraps.

send out scouts to check out potential real estate, and they vote on the best spots they can find. If your luxurious house gets to the top of the list, you get free bees!

The most successful real estate appears to have the following attributes:

- **space for the colony — at least six to eight frames/bars in volume**
- **small, defendable entrance (around 3 to 4 centimetres in diameter)**
- **secure, well-insulated space**
- **up high — around 2 to 3 metres above the ground is ideal**
- **smell of lemongrass oil gently wafting out of the box.**

The good news is that you don't need to build anything fancy; the nucleus box built for your hive type will work perfectly well. Just set it up somewhere off the ground, but ensure it is easy to reach if you need it quickly during inspections. On the roof of a shed or tied securely into a tree is ideal.

One golden rule for biosecurity: never try baiting the swarm trap with honey or old honeycomb, as this is a major disease risk. An old box and old frames that smell like 'home' is fine, provided you can be sure that they are disease free and have been well cleaned, with fresh foundation or wax top strips installed.

The best thing about a swarm trap is that you don't need to do anything other than check it every few days. If a swarm has moved in, there will be bees constantly coming and going at the entrance as they set up their new home.

If you see an occasional bee visiting, this is great news — they are scouts, measuring up the hive and reporting back. Don't disturb them, just get the camera ready. One scout will tell others, and you might find that there is a steady build-up of bees coming and going until, with a roar, the swarm arrives. This is your chance to get a spectacular video as the swarm moves in. The swirling, roaring mass of flying bees is an amazing part of nature, and something every beekeeper should experience.

The other advantageous role that a spare nucleus full of frames serves is as a source of spare parts. Every season there is a frame that breaks or has become too clogged with bee silk to use, and having a set of clean fames set up nearby and ready to go is incredibly handy.

Measuring up the drapes

As mentioned earlier, when the scouts first enter the swarm trap they pace out the internal cavity to make sure it is big enough for the colony.[1] So size does matter, and a trap that is too small will be ignored. You want it to be about the size of a nucleus hive, up to the dimensions of a single standard hive box. While swarms will move into a larger space, unless it is set up right in your apiary you will probably want to move it once it's occupied, so you don't want this too big. Another key consideration is that it has to take the same type of frames that go into your hive. It is worse than useless to collect a swam onto Langstroth frames that will never fit into your Warre hive, and vice versa.

Another key design feature is the entrance closer. Once you have the swarm in the box, you need to shut them in for a day or so until the queen starts laying eggs. Once there is brood in the comb they will 'lock' onto it, but until then they can sometimes do a runner. It is essential that they have enough ventilation to breathe, but that they still cannot get out.

Position, position, position!

Fundamentally this is about creating the perfect piece of bee real estate, and position is everything. You need as many scout bees as possible to find your trap. Honeybees evolved to live mostly in tree hollows, and the best sites are up high to give some protection from bears and other predators. A spot between 2 and 3 metres above ground seems to work best, but higher tends to make it hard to reach. Do not forget that the plan is for you to take this down when it is full of bees! I use the roof of my garage every year, and it works quite well — it is

Final bit of a swarm marching into the box.

about 2 metres high and gets morning sun, warming the hive. It is also convenient to reach to replace scent lures and any damaged or weathered boxes.

Another consideration is how easy the trap is to observe. When a swarm has moved in you want to be able to see the activity at the entrance without having to climb something or get too close. However, you also don't want it to be too easy to observe, especially from the street, or it may well get stolen!

I always have at least one swarm trap set up at each apiary, of the same type of hive as the others in the group. In addition to the chance of catching swarms from the wider landscape, it means that if I do need to split a hive I have a nucleus box right there ready to go, as when you need one you often need one quickly!

Scent lures

While the scout bees will check out lots of different nooks and crannies as they search for the perfect place to set up shop, their nose will really draw them to your bait hive. In the past, old honeycomb that had been in a hive for a long time was used and was quite successful, but this also carries with it an unacceptable disease risk. These days I keep some old frames that are either sterilized or from completely disease-free colonies handy, and these have the same smell of 'home' for bees without the risk of disease.

You can further load the dice by adding scent lures to the hive. Lemongrass is a well-known swarm attractant, and as a bonus it's easy to find and use. Just start cooking with it during spring (it is great in stir-fries), then keep the shavings and offcuts aside and throw these into your swarm trap. I also keep a small bottle of lemongrass oil handy when making up frames and put a few drops on the frames if they are going to go into a nucleus box or swarm trap. There is a wide variety of commercial mixtures that you can buy, and these are more expensive, but they definitely work!

In addition to the prohibition on the use of honeycomb as an attractant, there are a few other no-nos that you need to be aware of. While putting out honey might superficially seem to be a good idea, it is a major disease risk, so don't do it. Likewise sugar syrup or fondant — these will draw ants and other pests. Remember that the swarm carries food with it to help get established, and will work frantically to build resources. Food is not the attraction; space and security are.

Timing

The nice thing about swarm traps is that they are basically set and forget — a swarm either moves in on its own, or it doesn't. Their low maintenance nature means that I start setting them out from early spring when the colonies are active but before swarming starts. I take them back in for storage once I am sure that no further swarming is on the cards, generally in late summer or early autumn. I commonly use smaller nucleus boxes, but they can still keep a colony for at least a few weeks before they need a bigger box, so there is no desperate need to continually check the boxes.

The scent lures for the swarm boxes generally last a few weeks, and this is a good schedule to get into when checking the traps — you will either see bees working from the entrance or not. No bees, then just open up, clear out any spiders and replace the scent lure, and hopefully you will have better luck next time.

CATCHING A SWARM

Sometimes a swarm trap doesn't work, or the bees don't like the box and the swarm moves on until the queen can fly no further. The swarm will then congregate on a bush, wall, car, trampoline (no, really), or almost any other object and send out scouts to find their permanent home. This is the form that you often see on news reports, and often with a sensationalized treatment. The reality is that a swarm in this stage is directing all its energy to finding a new home. Bees are defensive, not aggressive, and if handled gently can basically be picked up with bare hands most of the time. Wear a suit and gloves just in case, though, because after a few days out in the cold their mood can become more desperate.

Whatever happens, always wear a suit and gloves when first approaching a swarm. You will sometimes see people catching swarms without protective gear, but they have

Roadside swarm collection.

usually carefully assessed the swarm off-camera before taking their gear off for a bit of 'stunt' beekeeping.

Also be sure to keep curious onlookers back, or at least warn them of the danger before proceeding. While a swarm might be perfectly happy to have people close at the start, this can change when they start to be moved into the box. Always take a safety-first approach.

Collecting into a nucleus box

The simplest and most direct approach is to pick up the swarm and place it directly into a hive box. Sounds simple, and often it pretty much is. This approach works well when the bees are close to the ground clustered in a single group on a tree, bush or structure. Collecting them when they are in this form is pretty simple: remove most of the frames to create space, place the box under the swarm (ideally with most of the bees in the box), shake

About to shake the branch to drop in the swarm.

Gently sliding the lid closed on the swarm.

the branch/chair/post they are clustered on and they will pour like water into the hive box.

Once most of the swarm is in the box, gently slide the frames into the box and place the lid on, carefully checking to make sure you aren't crushing any bees. Watch the bees at the entrance. If they are starting to accept their new home, you should see them stick their tails in the air and open a small white gland at the base of their tails. This releases Nasonov pheromone, which is a homing beacon for any stragglers.

Once most of the cluster is in the box, place it slightly off to the side of where they were clustered, and watch for the remaining bees to move into the box. If they start to re-cluster on the branch, carefully shake this onto a spare frame or section of cardboard and move the bees to the front of the nucleus box, watching carefully for the queen. She will usually be in the middle of the cluster, but not always! The remaining bees will try to cluster around her to protect her if she is not with the group, so this can be a clue. It is absolutely critical that you keep the queen with the swarm, as the swarm will have no chance of raising a new queen if she is lost or crushed. Likewise, do everything you can to save every single bee — they are carrying honey and will form the workforce for the new colony.

Once you have them all in the box, close the lid, pick them up and walk off. It's that easy! I like to keep them locked into a well-ventilated box for at least a day to make sure that they start building honeycomb to 'lock' onto that box and guarantee they will stay.

Letting the contrast between the white sheet and the dark box usher in the bees.

Collecting onto a white sheet

Sometimes the bees are up high or on an awkward spot, and this is where the white sheet method really comes into its own. It works by exploiting the fact that the bees are looking for somewhere dark and indoors to cluster, and we can use a white sheet to 'herd' them in the right direction.

I always carry a single bed white sheet with me when swarm collecting, as it is a surprisingly handy piece of gear. You want something that is light and breathable (no need for high thread-count Egyptian cotton here). Hold it against your mouth, and you should be able to still easily breathe through it — ventilation is important. Also make sure it is something that won't be missed from home — it is going to get stretched, torn, and generally abused as you catch swarms. My partner never noticed the sheet that I swiped from the linen cupboard, so it was obviously expendable.

Spread the sheet on the ground directly under the swarm cluster, with the nucleus box off to the side a little. The basic idea is that when you have to knock a swarm down from a height, or when you need to move them from an awkward space, they land on the sheet. From there they will be seeking the dark of an enclosure, which is hopefully the nearby nucleus box. It is worth having some pieces of flat cardboard or good gloves to help things along — bees are communal, so it's a good idea to place as much of the swarm as possible at the front doorstep of their new home, and then they should start signalling to the rest of the swarm to come join them.

Once they are all (or mostly) in the box the sheet becomes quite handy. Just pick up the edges and bundle the whole thing up like a big package, gently tying the corners together. This is where having a light, breathable sheet comes into its own — the bees can easily breathe through the sheet, and it allows you to collect stragglers without having to get absolutely every bee in the box, speeding up the process a lot. Once you have the box set up in their new position that night, the stragglers will be able to walk in.

Catching with a swarm bag

A few years ago I heard about an old method that beekeepers used to use: catching swarms into a sack or bag and then using this to move the swarm to their main apiary. There was even a fantastic story about one beekeeper who enveloped the swarm in her dress, carrying it home that way. I have terrible legs, skinny and white (classic Irish heritage), so dresses aren't my style, but using a swarm bag is an old trick that deserves to be

Example of a swarm that can be collected with a swarm bag.

Gotcha! Swarm in the bag, about to be tipped into the waiting Warre nucleus box.

used more frequently.

This approach suits a very specific situation, ideally where the swarm has collected on the end of a branch or other slender object, where they can be easily enveloped.

Using a swarm bag exploits the fact that once the bees are settled and waiting for the scouts to get back from looking for their new digs, a swarm will tend to cluster quite tightly together and not really have too many bees flying around. If you are quick, you can envelop them in a light cloth bag, gather up the top, snip off the branch and be on your way in less than a minute. Provided they are all in one group, you are guaranteed to have the queen as she will be in the main mass of bees (she is never out on her own).

This approach is incredibly useful for swarms in public places, where the disturbance of knocking them into a box is likely to send bees flying near people. The scouts and stragglers will quickly disperse back to their home hive once they realize that the main swarm is gone, and I once managed to remove a swarm from a tree on the edge of a children's playground within 5 minutes of arriving — a much safer option.

The downside of a swarm bag is that the bees can't stay in it for long. Even with the lightest cloth bag, the bees will overheat quite quickly, and that pisses them off, so this is a technique where speed is the most important factor. You grab, you dash, and then pour them into a more permanent hive (wearing gear!) at your apiary as soon as you possibly can. This brings us to the other downside of the swarm bag: bees really aren't that different to us in many ways, and having some weirdo sneak up on them and kidnap them in a sack before running off tends to piss them off no end! They will be a bit aggrieved at this treatment, so be prepared for some crankiness when you do release them. Just be as gentle as you can at the other end, placing the bag mostly in the hive box that you want them to move into before turning it inside out. Once they are mostly in the box, give the bag some gentle shakes to dislodge the rest and close up pronto!

The actual swarm bag itself can be almost any light cloth, provided it is a fine enough mesh that the bees cannot get through it. I have a pretty simple test to see if a bag is suitable — hold it against your face and see if you can easily breathe through it. If you can, it should be fine. If breathing is a strain at all, then the cloth is probably not open enough. In these environmentally conscious times there is a wide range of cloth grocery bags and many of these are a fine mesh that is perfect for this type of swarm catching. In terms of volume, these are cheap, so it is worth having a few. I have a large one that once held a dog's bed; it is about 20 litres in volume so is more than enough for even large swarms. I have a few smaller ones, about the volume of a pillowcase, and these work pretty well for most situations.

It's important that the bag has a wide mouth — you want to be able to envelop the whole swarm without touching it, so wider is better. A drawstring close is also nice, but not essential. It does make it easier to tie off the top of the bag, though, and also gives you a handy carry point. The drawstring that I have lets me hang up the bag in the car, making it easier to give the bees airflow as I head to their new home.

HIVE REMOVAL FROM STRUCTURES

The most difficult removals are where the bees have already started setting up a hive, building comb and raising brood. This is their home now, and your job in this case is less to catch a swarm and more to move an established hive into a more appropriate residence. The one saving grace of this approach is that there will normally be brood present, which allows the beekeeper to inspect for disease as the removal is in progress. You also tend to score a bit

Dermot carefully cutting through to the colony.

Tree hollow opened; now to extract the bees!

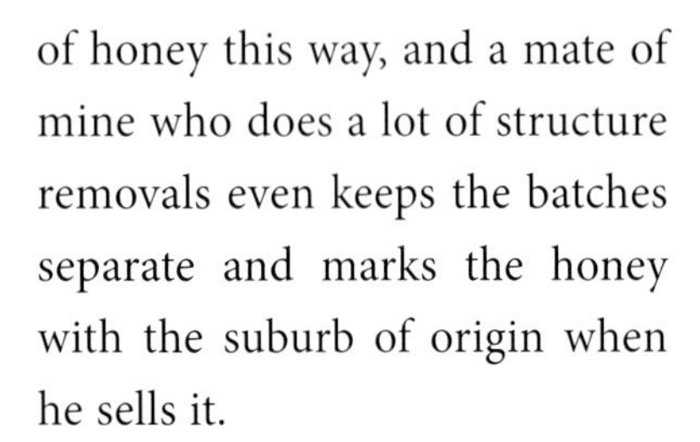

of honey this way, and a mate of mine who does a lot of structure removals even keeps the batches separate and marks the honey with the suburb of origin when he sells it.

Removing an established colony like this is part beekeeping, part building demolition and part puzzle solving. It is also a brilliant way to get a well-established colony with good local genetics — sometimes these colonies have lived in the house/chimney/shed for years or even decades. They have stood the test of time and these tough, streetwise bees can be fantastic to look after.

However, getting to them and then getting them out can be tricky and quite intense, hot and painful work. A friend of mine who does a lot of this uses a combination of thermal imaging cameras and a stethoscope to figure out where the main body of the colony is, and then dismantles whatever is in the way to expose the honeycomb for removal. It is then a race against time getting the brood into a box before it is chilled.

Getting access to the colony can be a little terrifying. It is rare that the bees are in a

Bee vacuum in action. Yes, it is an actual thing. (Credit: Dermot Asls Sha'non)

Lifting the bees out of the vacuum unit to place in a hive box. (Credit: Dermot Asls Sha'non)

simple, easy-to-get-to area. Typically, they will have holed up in a wall or deep in a log, and a combination of crowbars, reciprocating saws and even chainsaws often has to be used to get to them. Using a chainsaw in a full bee suit is not for the faint hearted, and you need full safety gear on (helmet, ear protection, gloves, chaps as a minimum), as so many things can go wrong.

'What about the bees?' I hear you ask. My colleague has another ingenious bit of kit for this: a bee vacuum. Seriously, I am not making this up — this is a common piece of kit for such specialist beekeepers, and it is surprisingly gentle on the bees. A commercial vacuum cleaner is rigged up to a sealed hive box. The bees are vacuumed up, but the pressure drops once the air is moving through the larger box, allowing the bees to grab hold of the frames and hang out. A filter at the exit to the box means that the bees aren't sucked into the machine itself, and this system allows the adult bees to be vacuumed up into a hive, removing the need to smoke them. Once the bees have been vacuumed off the honeycomb, it is cut into sections; the brood is carefully removed and placed onto frames with rubber bands. This becomes the new brood box, and the box of frames that the bees have been vacuumed up into is then added to this box to establish a new colony.

The honey is cut out into raw chunks and placed into tubs for extraction to be crushed out from the honeycomb later. This is a really important point — you want to completely remove the honey and not try to keep it with the cut-out colony. Cut comb will leak profusely and will drown the bees. With the honeycomb cut out, they have nowhere to store it until they build new comb. Even worse, if the honey gets into the lines of the vacuum, it will make the bees' wings stick to the walls of the vacuum system, shredding them and basically killing them. Broken honeycomb is also a magnet for robber bees and pests like Small Hive beetle, so you want it out of the mix and in a separate tub as soon as possible.

SWARM KIT

With swarms, time really is of the essence. You need to get to them fast, while they are still deciding where they might like to live. If you wait until you get the call to assemble what you need to catch them, then the time you waste may well mean the difference between getting a new colony in the box or watching them disappear off into the sunset.

So you need a 'grab and go' kit that is prepared well ahead of time. During peak swarm season, I even keep a basic kit in the boot of my car, as I will often get a call towards the end of the day. I can just duck over on the way home, grab the swarm and be back in time to cook dinner. Occasionally my partner will notice me ducking out to check on the swarm siting on the garage roof, but mostly it is the perfect crime!

This is my recommended kit:

- **nucleus hive with frames (ideally 50 per cent foundation, 50 per cent foundationless)**
- **white sheet**
- **swarm bag**
- **spare suit and gloves (often just a veil or half suit)**
- **swarm lures (sachet of scent to attract bees)**
- **bee brush**
- **sheet of cardboard (handy to scrape bees off walls, or to brush onto)**
- **water mister**
- **sugar block and pollen patty (for the bees to snack on, not you)**
- **sting relief spray**
- **bee repellant spray (to get them out of tight spaces)**
- **disposable gloves.**

Sometimes they stay on the branch, creating a beautiful open-air hive.

This sounds like a lot, but it is actually pretty cheap, compact and easy to put together and should fit into a small box. I even use an archive box, as in a pinch you can easily fit an extra swarm into this box as well. This is handy, as at least once per season I will get a call-out where there are multiple swarms. One of my colleagues was even getting a second swarm in the box when a third one turned up!

If you are serious about swarm collecting, it pays to have a bunch of nucleus boxes ready to go by spring, ready to expand your apiary. I tend to map out how many swarms I want to catch at the beginning of spring (usually a certain number of new colonies for my apiaries) and once I have hit this then additional swarms will be joined to colonies to strengthen them.

11

It's not about the honey

Creating visitor experiences and more

Watching someone try pure, single-source honey after a lifetime of bland supermarket honey is one of the great pleasures of beekeeping. Showing the public bees, honey and even artistic events centered around bees is where we connect with others, broadening horizons and sharing ideas.

Being the beekeeper for a building or an organization instantly connects you with dozens and even hundreds of people from all walks of life, most of whom will have had only a passing experience with bees. Even keeping bees in your backyard recreationally will instantly tag you as 'the bee guy/gal' among your friends and family. Apart from fielding endless questions about bees and beekeeping, one of the constant themes that comes up is what to do about saving bees. This drives some bee folks crazy, but I really like it — being a beekeeper gives you a practical insight into what makes gardens and city landscapes more sustainable, and people are really interested in doing something practical to help. Best of all, there are lots of ways that we can make genuine lasting changes to how we live that

Honey, chocolate and a beautiful garden. Lovely.

Fresh honeycomb from the garden and a cheese board — the Italian Embassy for World Bee Day.

will improve sustainability.

I think the reason that people turn to beekeepers is that we have a practical connection to the landscape. As I talked about earlier, the change in perspective that becoming a beekeeper gives you is unique, and is something I have become passionate about sharing. When family, friends and colleagues ask you questions and show interest, it is an opportunity to spread a little bit of awareness and interest, supporting another step on the long road to a truly sustainable city.

There are lots of ways to connect with people, but I have found that practical, visceral experiences have the most lasting impact. Visitors come away having had an experience they will remember for a long time and will talk about with friends. At the same time, hosting an event with hundreds or even thousands of people can be a daunting prospect, so I wanted to share some of my experiences, tips and tricks to help you create awesome experiences.

FOOD

There is that old saying that the way to a man's heart is through his stomach, and the good news is that this doesn't just work for guys! Honey was the original reason that humans came into contact with bees, and as far back as ancient Egypt we realized that supplementary pollination by managed bee colonies improves the yield and quality of crops. The one downside to emphasizing honey is that it is really only produced in commercially viable quantities by one species, the European honeybee. This has led to a complete dominance of honeybee images in the media, with sometimes little attention paid to the 20,000 other bee species. Despite this, honey remains a powerful way to connect people with their landscape, and you shouldn't pass up this opportunity. Plus, it is really fun.

The original office honey that started it all for me.

Honey for the homeless

After establishing the Austrian Embassy apiary in 2021, the current ambassador, Lukas Strohmayer, quickly found that it became a family affair, with his daughter and son turning out to be enthusiastic and accomplished beekeepers. It would have been easy to just put the honey in jars and give it out as gifts, but the productivity of the beehives and the ambassador's social conscience led him to try something completely different. Thus was born the most innovative and delicious bee-themed charity event I have ever been involved in — the Friends of the Apple Strudel!

Friends of the Apple Strudel is a series of events where embassies showcase their famous national dish, then auction off a sample cooked by an expert chef, as well as honey from the beehives of several embassies. The funds go to support Canberra My Home, a Rotary charity initiative to create housing for the mentally ill.

Following the initial event, other embassies enthusiastically answered the call — nothing like a bit of healthy competition to get people to step up! In addition to other ambassadors, the events attracted local and federal politicians, community leaders and of course bee enthusiasts, and have already raised thousands of dollars to support the charity.

There are a few elements that made this such a successful initiative, and they are worth considering if you plan something similar:

The Austrian ambassador, HE Lukas Strohmayer, and his daughter Katharina.

The Austrian Embassy hives.

Apple strudel made with embassy honey.

Genuine connections: The ambassador has had a long-term association with Rotary and social justice causes, resulting in a deep understanding of what they are trying to achieve and who needs to be involved.

Local food: Sourcing the key ingredients locally (and really locally — you can see the hives from where the strudel is served) brings a genuine connection to the landscape, plus a curiosity around what their honey tastes like compared to the other embassies!

Fun, playful competition: Nothing like a bit of rivalry to garner interest, and rotating the events around different embassies maintains interest.

This is a beautiful way to bring together people to share food, conversation and support for a worthy cause — and hopefully an inspiration for you to include charities in your events.

Honey sales are the obvious way people connect with the bees living on top of their workplace, and the influence of this should not be discounted. The first honey 'brand' that I ever created was for my workplace at the time, and this quickly became a talking point within the company, as well as creating some unique gifts for clients. A sense of place and belonging is one of the most powerful influences for human beings across the planet, and honey from their building will have a unique flavour profile. This is one way that the general public can start to see the world through the eyes of a beekeeper, helping them to realize how connected even urban landscapes are to food production. It also opens up important conversations about green spaces, pollinator-friendly street trees and the use of pesticides.

The really magical thing about honey is that it is unique for each location, and can even be specific to each colony in an apiary. Honeybees have a behaviour called floral fidelity, where the scouts will find a large source of flowers and then direct the entire workforce to collect from those flowers until they are finished. The result is that honey from this hive will be dominated by the largest source of flowers, and some beekeepers can even make honey purely from one source of flowers. In cities there is normally a diversity of flowers blooming, so each apiary within the city will feature a unique blend of floral sources, leading to some really complex flavors in the honey that are like a fingerprint.

This provides an opportunity for local branding of honey, and in fact some urban beekeeping businesses have done just that. My own experience is that this is a very powerful

There is a wonderful variance in the appearance and flavour of honey.

A trio of honey-themed treats created in the Parliament kitchens.

Hard to describe how satisfying it is to have your own honey label!

way to connect with people who would not normally think about sustainability in cities. We have traditionally seen our cities as nature-free zones, devoid of life and certainly not suitable for boutique food production. Creating delicious, region-specific premium food subverts this idea.

The journey doesn't need to stop at the honey. There is a world of specialty urban food opening up out there, from herbs to vegetables and even food forests bringing flavour and fun into our urban areas. Not only are bees (especially native ones) critical to production of these crops, this opens up recipes that pair with local honey, creating ever more complex interactions. The chefs at Parliament House have turned this into a minor art form, creating unique desserts for official dinners. For my pollination hives the restaurants that receive the fruits of the bees' labour have found creative ways to pair the honey that highlight the local nature of the production. However, my favourite story is about how my honey brand got its name.

I had been supplying surplus honey to my local butcher for a few years when he decided to highlight the locally produced nature of the meat he was selling. Bede is a brilliant butcher — he specializes in single-source meat from local farms and has been using my honey in glazes and sausages for years. When he was showcasing the local producers on the daily specials board, he described my honey as from 'the mean streets of Evatt' as a gag (our streets are decidedly middle class). He immediately got a bunch of complaints from residents outraged that he had made their suburb out to be rough and tough. Bede and I both thought this was hilarious, and Mean Streets Urban Honey was born!

HONEY INFUSED DRINKS: RELEASING YOUR INNER VIKING

Once we had started producing gift honey jars at Parliament, we hit upon a unique problem. Many countries do not allow the importation of honey due to biosecurity, meaning many of those who received our honey could not take their gift home. Even within Australia, it is illegal to move honey between some states, and this is critical to maintaining biosecurity so it is not a rule that we wanted to break.

At the same time, we hit upon a problem when packing down some hives at the end of autumn: some of the honey left in the frames was unripe and about to start fermenting. Putting this in jars would result in exploding gifts — not the best idea, especially for foreign dignitaries! One of the things that judo teaches you is to go with the flow rather than trying to fight an irresistible force, so rather than trying to resist fermentation we provided the honey to a master mead maker who could guide it to a less explosive and more delicious destination!

While some countries will not allow imports of honey, most allow alcoholic drinks, and so the inaugural vintage of the Parliament mead was born, later to be followed by honey-infused vodka!

Mead is an ancient form of alcohol that was particularly popular with the Vikings, and at its most basic simply involves adding water to honey and letting the natural yeasts within the pollen take over and produce alcohol. This can also produce undrinkable swill, so modern mead makers have turned this into a fine art with different yeasts, herbs and even fruit additives. Meads are incredibly diverse in flavour, but at their heart always feature honey as the source of fermentable sugars. This gives the mead the

Parliament honey-infused alcohol — of course!

unique flavours of the honey, especially if you use a single source.

The experiment with mead was a success, and the first run was featured at the second World Bee Day held at Parliament, selling out almost immediately. People seemed to really love the idea of an ancient drink that was infused with local honey, and best of all, it meant there was a form of Parliament honey as a gift that could be taken home by every potential visitor. This started the next evolution in the initiative, with both mead and honey-infused vodka eventually added to the gift shop.

While this was a lot of fun, especially when we could share these drinks at official events, our enthusiasm was tempered somewhat by the widespread damage alcohol does in society. I have more than a few friends who won't drink for personal reasons, including some recovered alcoholics. As a result, we went for premium products that are inherently low volume, and made sure that alcohol was not the main focus. Alcohol is so pervasive in our society it is really easy to let it just become infused through everything, so it is important to check yourself every now and then and make sure that the (slightly tipsy) tail isn't wagging the dog.

ARTWORK

The art of bees and beekeeping has inspired us for centuries, so it should have been no surprise how much art would feature in my beekeeping journey, but it was. Probably because I have no discernable artistic talent myself! What I do have is the good fortune to meet many talented artists as part of my work, and the good sense to listen if they are keen to try something with bees!

Beauty and art coming together to celebrate Indigenous culture.

Honeybee democracy: Celebrating the history of Parliamentary beekeeping

The entrance to the exhibition.

Being included in a museum exhibition next to historical figures was surreal.

You know you are getting old when you are literally a museum exhibit, but when the Museum of Australian Democracy came to me with an idea for a bee-themed exhibit I was never going to say no! The museum exists as a living social and political history, located in the original Parliament House in Canberra. Most people think the honeybee queen is in charge, but in reality the colony makes decisions by voting, and they represent the oldest and purest form of democratic government on the planet, literally voting to decide life-or-death decisions.

As it happens, Old Parliament House, where the museum is located, is also where the first Parliamentary beekeeper kept his hives. Quite the larrikin, MP William Yates asked the Speaker of the House, Billy Snedden, if he could keep his bees at Parliament on 1 April 1976. Assuming it was an April Fool's joke, the Speaker agreed. Except that it wasn't a joke – the bees arrived shortly after and spent many years pollinating the gardens and producing honey for gifts.

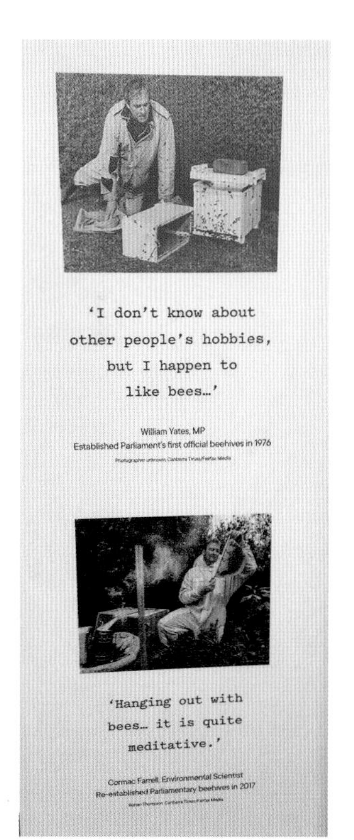

Out of this initial conversation came a whimsical, beautiful exhibition, still running today, that brings together the history and current practice of beekeeping in the Parliamentary precinct. In addition, there is a strong interactive element, with visitors recording their thoughts and reflections on honeycomb panels that then adorned the walls. I also spent a few hours in a recording booth talking about bees, beekeeping and how bees manage the voting process, to provide a narrative as people walk through the exhibition.

Having bees at the centre of a museum exhibit is a pretty unique

experience, but for me the highlight was the opening of the exhibit, where I got to meet the family of the original Parliament beekeeper and share a meal with them. The thing I found most interesting was how they had all gone on to blaze trails in their respective careers, innovating and changing the way things were done. Several are beekeepers, but they all have questioned the way things were done and sought to change things for the better, often with an environmental focus.

It was a chance for lots of bee puns, a fun panel discussion during the official launch and then later lots of discussion around the parallels between our democratic institutions. Happily, even the toughest parliaments tend to be kinder than some of the brutal practices within a honeybee hive! HiveMind, as the exhibition is called, is also a great example of how you can find yourself collaborating with people you had never expected and can be richer for the experience.

Old and new gear from the Parliament apiary on display.

My suit and smoker from the modern Parliament apiary.

Messages from the public, inspired by bees, became part of the exhibit.

Old and new together.

Nature's original artists.

In one way all beekeepers are consumers of artwork, in the form of our jar labels. I have yet to meet someone who just plonks text on the outside of their jars.

I have found a few principles handy when working with artists:

Pay them. You can't pay your rent with 'exposure' and neither can they. I have always insisted on paying artists for their work and have never regretted it.

Be inspired by their past work. You might see a really interesting idea that you can explore with them, and it also means you can show to your sponsors and partners what sort of art they do.

Talk them up. Once the job is done, make sure you highlight their work to anyone who will listen. Especially for artists starting out, word of mouth is everything!

Consult widely. For your corporate clients, community groups and especially Indigenous artists, there may be permissions or established relationships that you have to manage carefully. Be patient, respectful and ask lots of questions.

For all the planning in the world, it can also be the surprises that can have the most meaning for you. Working with artists gets you and your ideas out there, and this sometimes comes back to you in a good way. An absolute highlight of the Parliament apiary was being

The spectacular Lego Parliament House, now on permanent display.

called in to see a pre-showing of the Lego Parliament model, created by the famous artist, The Brickman. It had captured all sorts of weird characters around the building, including myself. Probably only a moment's thought for the artist but for someone like me who loves Lego, working in a Lego-mad engineering company, becoming a figure in an official installation was one of the best days of my life.

TOURS

As beekeepers, the relationship between the landscape, our native bees and our honeybees is the most interesting and tricky management puzzle to figure out. Showing people directly is the best way I have found to explain this. Whether it be the diverse inhabitants of native bee hotels or the frenetic activity of a honeybee entrance, people can't help but be fascinated.

During the middle of the pandemic, we

Garden tours were the perfect Covid-safe event. (Credit: 5 Foot Photography)

Showing staff the inner workings of the Parliament hives. (Credit: 5 Foot Photography)

Showing off the strange, alien brood chamber of a stingless bee colony.

still wanted to give people a chance to experience the connection of bees and the surrounding gardens, even if we couldn't do indoor events like honey tasting. Outdoor events were much safer, and just as popular. It actually turned out much better than we had expected. Most botanists and horticulturalists are nerds, and we know it. We rarely expect the wider public to care as much as we do, but when you are passionate about something it tends to hold people's attention. When it is something that is fascinating in itself, like the interplay between flowers and pollinators, you have a winner.

It doesn't have to just be honeybees, either. While a honeybee hive is spectacular, standing in the working stream of bees is fairly hazardous, akin to standing in the middle of a busy corridor of Parliament when the division bells are ringing (seriously, they run like crazy to get to the vote). Eventually someone is going to get pissed off at you.

Safety always has to come first, and this is where native bees really shine! Even the natives that have a sting almost never use it, and then only if someone grabs hold of them (hot tip: don't do this). People can get really close to native bee hotels and then go home and put their own together in their garden the same day. So many people have told me after an event that they were inspired to put a few bee hotels in their garden, and then shared their delight as the native bees started to move in.

Honeybee tours take a bit more logistical organization, but are spectacular and fun. You need protective suits, first aid supplies (just in case), training and proper insurance. The payoff is that the hives themselves are endlessly fascinating; I love it when people who are really scared of bees sit close to the hives and start to realize how gentle and interesting bees are. One of the best events was run by the Embassy of Sweden during the first UN World Bee Day, and just involved putting people in a suit and letting them sit by the hive entrance, watching the bees work. A simple concept, but it turned out to be one of the most popular activities; you could see people's initial nervousness dissipate, to be replaced by fascination.

CITIZEN SCIENCE

A super-cute Leafcutter bee (*Megachile maculariformis*).

The community concern for the plight of bees has translated into a desire to do something practical that genuinely adds to conservation, and taking part in the many excellent citizen science initiatives out there is one of the best things you can do. They don't take much gear, are usually already set up, and can cater for very large groups of people — after all, the whole idea of this is to bring in as many people as possible.

While the media tends to fixate on honeybees, the reality is that the other 20,000 bee species need our attention a lot more. At a basic level, we still don't know that much about how many native bees we have across the landscape, and how their populations change year to year. We know there are lot of them, and that they are incredibly important, but most of the information we collect on bees is from managed honeybee colonies. This is still important for agricultural research, but the reality is that we monitor these bees because they are worth money to us.

Recent research has also shown that honeybees, especially managed honeybees living in commercial hives, are not a good surrogate for native bee populations. This is a problem in that honeybees have been extensively studied as a 'model species', often used as a surrogate for all bees. While some of the things that stress native bee populations also impact honeybee colonies (e.g. pesticides, habitat loss), there are major differences, enough that we simply cannot rely on monitoring commercial honeybees and assume that this translates to our native species.[1]

It would be nice if there was enough funding to pay researchers to collect data on all the critical aspects of the environment, but the reality is that there is never enough to go around (reasons for this could form a whole other book). However, the skill in most ecological research is in the design of the survey and then in drawing insights from the collected data. The fieldwork of data collection is often much simpler, and has a huge added bonus: it is the fun bit. Leveraging citizen scientists to collect large amounts of data across the entire landscape is an increasingly effective way to stretch scant research dollars and build interest in the results at the same time.

Megachile erythropyga **entering a bee hotel. (Credit: Dr Kit Prendergast)**

It is important to understand that not all citizen science programs are created equal. Some are limited in how useful they can be by poor design, inadequate expert support and unclear objectives. The best ones are fun, but still provide a useful dataset for researchers to draw on. The best citizen science projects have a few features in common:

- **They are designed and run by academics with deep expertise in the field.**
- **They use a simple method that does not require complex equipment.**
- **They have clear objectives and limitations on what they are trying to achieve.**

One of the best programs, which I love taking part in, is the Wild Pollinator Count (https://wildpollinatorcount.com/). This happens twice per year in spring and autumn and gathers data on the abundance of pollinators by simply observing a flower (or a group of flowers) for a set period of time and recording who visits. The Wild Pollinator Count has been running since 2014, collecting valuable data on native bees, wasps, ants, beetles, butterflies and the myriad other pollinators that visit plants.

Rather than starting your own project, have a look and see what is happening out there in your area, and do what you can to get people involved. Your native pollinators and their researchers will thank you!

HOW TO BEE ONLINE: SOCIAL MEDIA AND FINDING YOUR 'BRAND'

Social media has been a phenomenon in bee conservation, but at the same time it is a double-edged sword. It is a powerful tool for informing, connecting and galvanizing action within the community to protect and enhance bee habitat. At the same time, it can be overwhelming and toxic, particularly with the 'pile-on' nature of online groups when they turn sour.

Social media is just about made for bees. The beautiful colours of native bees, the spectacular swarming of honeybees, and the endless fascination of honey in food combines to create endless, unique content. It can also be an invaluable resource — every year I learn new techniques and insights from watching other beekeepers do their thing. The beekeepers themselves are also pretty unique, beekeeping basically being where farming and crazy people meet!

One attitude that you will run into a lot is a concern that the overwhelming focus on honeybees in the media is crowding out the attention that should be going towards conservation of native species. This is definitely a real issue: honeybees generate huge interest in all forms of media, but other bees are almost completely absent, even if many of the issues are common to many species. Are honeybees acting as ambassadors to raise awareness of wider pollinator conservation, or are they just hogging the limelight? Some researchers at the University of Pennsylvania specifically looked

A hive roof full of honey.

at this, and found that honeybees act as a familiar place to start for many people, and make it easier to draw connections to their less well-known native cousins.[2]

The flip side of this also highlights a danger: some folks get so frustrated at the focus on honeybees that they try to dissuade people from being so concerned. The risk is that this can backfire, making people think that the issues with pollinator conservation have been 'solved' and that they can stop worrying. A better approach is to lean in to this interest in honeybees, and then use it as a mental anchor point for people to explore wider issues around urban ecology, landscape conservation and pesticide use. With the advent of social media, you have the potential to start long-running conversations with the community that can engage hundreds, even thousands, of people.

Let's look at four of the more dominant forms of social media.

Instagram

Primarily a photo- and video-sharing platform, Instagram has lots of beekeepers and beekeeping businesses, and you can link through to online shops. The focus on visual content really helps native bees, as their stunning colours shine, and there are also fun, innovative games like queen-spotting challenges. Instagram offers a generally nicer, more positive community online as well.

Facebook

One of the original social networks and still one of the most heavily used. The capacity to create a specific, members-only channel is a fantastic function for community gardens and bee clubs. The main disadvantage is the avalanche of conflicting advice and occasional toxic commentary, but Facebook is still the mainstay for organizing and promoting large events.

Twitter

With a focus on fast, short messages that drive engagement and discussion, Twitter was really fun. The ability in this style of social media to curate your feed of engagements can limit your outreach, but is a great way to minimize some of the less pleasant elements. You did need to be on it a lot, but for science, nature and beekeeping it allowed you to connect with experts from around the globe in seconds. At the time of writing Twitter was undergoing change under a new owner, and renamed X. In many ways it seems to be dying, which is a shame as it holds a lot of great memories for me. However, a range of similar networks

Hanging out with a swarm on a spring afternoon.

(Blue Sky, Threads, Hive) are springing up to fill the vacuum, so this style of social media will never really disappear.

YouTube

Long-form videos are the name of the game here. YouTube is a fantastic instructional resource for almost any skill, with beekeeping no exception. Lots of really great beekeepers provide in-depth advice on this platform, and it offers a great way to showcase processes like building native bee hotels and honey extraction.

When done right, the use of online branding and social media allows us beekeepers to become so much more than just the sum of our parts. Swapping information, ideas, contacts and techniques is a real boon for everyone, no matter where you are on the journey. Most of us credit our success early on to having a good mentor, but having a decent social media community around you is almost as good, and better in some respects.

Building a personal 'brand' was not something I ever really set out to do, but it does seem to have happened — I now regularly get questions on all sorts of topics. Sometimes I can even provide a coherent answer! An innate sense of curiosity, some basic manners, and a reasonable bullshit detector seem to have served me well in navigating online spaces.

Finding your purpose

When I was working in the private sector, we often had to take part in developing some form of mission statement for the company. I always hated these — they seemed a bit of a waste of time — but later on I started to appreciate them. Don't panic; I'm not suggesting you do a mission statement, or the insufferable workshops that go with them. But there is a bit of value hiding in all the corporate babble around these. It makes you think about what you are really trying to achieve, and whether that has value.

The Japanese have a really wonderful concept called *ikigai*, which translates to your purpose for living. It ideally contains three things:

- **something that brings you joy**
- **something society values**
- **something you are good at.**

Find something that fits all three at once, and you have found your reason to be on the planet, your *ikigai*. Within the bee world, there are so many specialized little jobs that it is pretty easy to find something you are good at and that brings you joy. I have a friend who is a honey sommelier. She specializes in knowing about all the different flavours of honey that are out there. For myself, I have become good at managing all sorts of different hives when most beekeepers only use the one type. This set me up beautifully to support embassies, as many wanted to run the hive from their home country.

Japanese garden, Eisho-ji Temple.

Bee diplomacy: Connecting countries through bees

A little more than a year after I started keeping bees at Parliament, I had a somewhat random message appear in my inbox. The Swiss ambassador was interested in having a chat about bees. Figuring that I would at least get some decent chocolate, I agreed on the spot. I did indeed get chocolate, and something much more valuable: friendships across a unique beekeeping community. It turned out that there were three ambassadors who were planning to keep bees but were not sure how to proceed. In the case of the Swiss ambassador, Pedro Zwahlen, he was far more experienced than I was, having kept bees commercially in his youth in Switzerland, but he was unfamiliar with Australian conditions. The Swedish ambassador, Par Ahlberger, wanted both native bees and honeybees to pollinate a special garden he was designing in honour of the famous botanist Solander. The Slovenian ambassador, Helena Drnovsek Zorko, had already started keeping bees at her embassy and Slovenia was petitioning the United Nations to establish World Bee Day.

These three pioneering ambassadors became the core of Canberra's diplomatic beekeeping community that has now grown to include Austria, Italy, France, Belgium, Slovakia and Portugal. Each embassy has taken to keeping bees in a way that reflects their culture, incorporating native bee hotels, unique hive designs, artwork and events.

The Swiss embassy sponsors film nights for World Bee Day. The Slovenian, Swedish and Belgian embassies run open days where people can visit the hives up close and learn about sustainability. The Austrians use their honey for charity events. The Slovakian embassy uses handcrafted national hive enclosures and sells honey to support Ukrainian beekeepers impacted by the Russian invasion. All different, yet all coming together to support each other and reach out to the wider community. The best bit has been that I haven't had to do that much — a bit of training, some friendly encouragement, and the sheer fun of bees soon took over. Along the way I had to learn how to use lots of different hive designs, which helped my beekeeping to no end. So take those small, random meetings over coffee — you never know what you might just start off!

Cutting the ribbon to open the inaugural World Bee Day event.

Ambassadors and beekeepers meeting the Governor-General.

Unveiling the Slovenian AZ hive on World Bee Day.

The Parliament gift shop now has an entire bee section.

More than just finding your purpose, finding your personal *ikigai* or mission to do with bees is important when defining your online persona: what makes you special? It also sets the boundaries of what you do and don't want to do, which is really important. People are going to ask you to get involved in all sorts of side projects along the way, and you need to figure out what fits with your style and purpose.

When I was starting out in beekeeping I found that I was pretty good at catching and rehoming swarms. Around the same time many of the 'old hands' had started to give swarm chasing away, but nobody had told the bees, so it was still a pretty huge job every spring and summer. I didn't really feel ready, but someone had to step up, and me and a few of the other beginners just threw ourselves into it. Plenty of mistakes were made, but we developed a lot of skills, fast. The community needed swarm collectors, it was a really fun part of beekeeping, and as it turned out, I was pretty good at it. A perfect fit. As a result, I always had extra colonies of bees handy, which helped when projects came up to establish new apiaries, and things just grew from there.

Saying no to things is also a critical skill. It is always presented as being a case of 'just say no', but the reality is way different. What if it isn't a hard no, but a 'not right now' situation? What if you don't want to sound like a jerk (or even worse, actually be a jerk)? Being able to figure out that something isn't going to fit into what you are doing allows you to give the most important answer when politely refusing: the honest one.

Photographs, the key to connecting with your audience

The biggest advantage in connecting with people and building your personal brand online is good photos. Native bees in particular are visually spectacular, coming in all shades of blue, red, green and yellow. Most online communication relies heavily on pictures and video to tell the story, and this is an opportunity to rebalance the focus towards native species in a way that intrigues and educates. Honeybees don't miss out, either. The whole process of honey extraction is extremely photogenic, and the frenetic activity of swarms and hive entrances is just made for online video.

Smartphones are an urban beekeeper's best friend, helping you stay in touch with colleagues and clients, record hive inspections, and provide links to your online hive monitors, all in the palm of your hand. But the camera is king. When choosing your next smartphone, really check out the camera, especially the macro function for close-ups, and then go nuts. Photograph everything. I find it is even worth having a checklist of 'must capture' moments.

Don't just focus on the finished product, either. Capture and share the entire process, from building a native bee hotel to checking honeybee hives, and then the final extraction of honey.

There are some classic 'must capture' shots that you should pay particular attention to:

Bees coming and going from native bee hotels and hives, lighting the smoker:

Native stingless bees guarding the entrance.

Lighting the smoker always looks great on film. (Credit: 5 Foot Photography)

Inspecting hives, especially the internal structure and how they are arranged:

Queen bee marked with a red dot of paint.

Honeypots of a native stingless beehive.

Honeybees stealing back their honey.

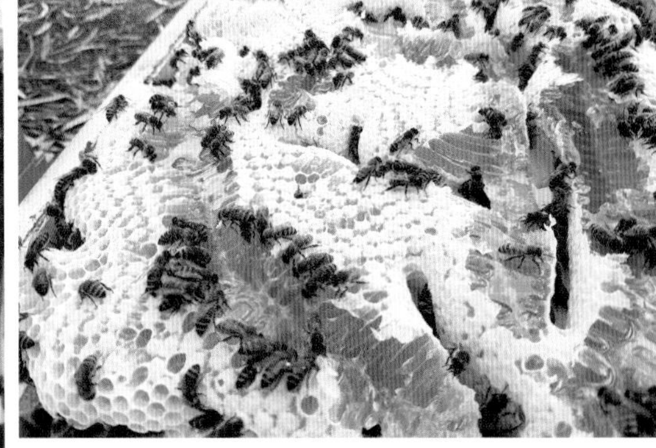

Event photos, where people are having fun interacting with the bees or the honey:

A range of honey flavours on offer.

Field day in the training apiary.

A beautiful frame of honey.

A friend on your shoulder.

Honey extraction, especially the first pour of the latest harvest, is a classic moment that you should always capture. You will also be producing honey that is constantly changing in both colour and texture, so be sure to keep a few jars for comparison.

Crushing honey out of the combs with a press.

Just a few honey jars!

The variation in honey colour between two seasons can be really stark.

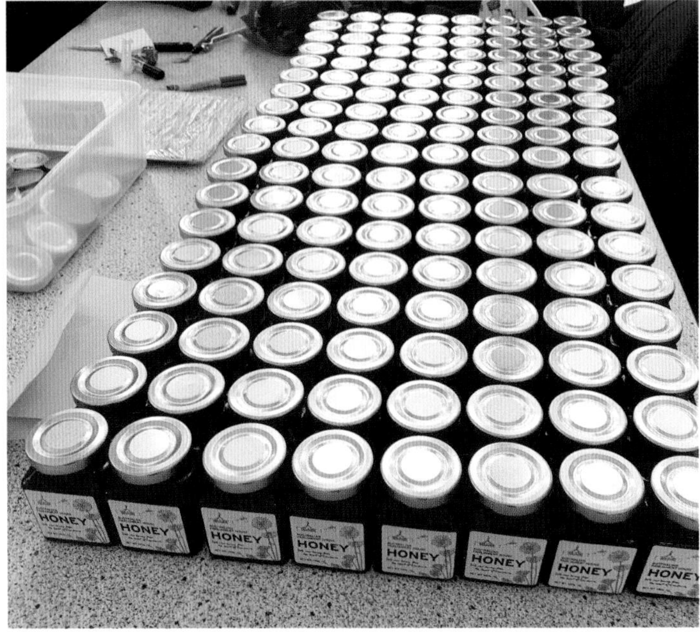

Being the better person online

We live in a world where being a white, straight male allows you to glide effortlessly through situations that are a major problem for women, people of colour and LGBTQI+ folks. Just reading that sentence will turn off some folks, and I get that, but bear with me. If you are a guy reading this, I want you to do something that only you can really do: be the good guy, and enforce that standard in others. I really became aware of this with female friends, as their experience online is dramatically different to mine, and not in a good way. There is a small (but significant) group of guys who are toxic jerks online, and if you let their nastiness slide by it makes online spaces hostile for a lot of awesome people.

Do	Don't
Bee positive, always. Even when highlighting a serious problem, offer a solution. **Focus on the issue, not the personality. Avoid the pile-on.** **Get permission to use photographs, always.**	Forget to share credit, especially when working with early career artists. Let personal attacks get to you — they say more about the attacker than you. Forget to block/mute trolls; life is too short for these jerks!

Here is the thing — there is something we can do that really does seem to make a difference. There is a great saying that the standard you walk past is the standard you accept. When you see trolls online, don't just walk past, take a few seconds to slap them down.

Remember that honey catches more flies than vinegar!

Check to see if it is an automated 'bot' account (e.g. the account has few followers and a basic profile image) and report it as spam. If it is a real person, tell them to pull their head in and apologize. I have started to do this, and it is amazing how a troll will argue endlessly with a female/gay/non-white person, but just whimper and melt away when there is an intervention from another white guy. It is basically a weird superpower, and one that we should use more.

Sarah is the most experienced beekeeper in the Parliament 'bee team'.

Case study:
The missing Cheerios bee

Too cute!

Cheerios is a honey-flavoured American breakfast cereal that really leans in on the whole honeybee thing, even down to the bee mascot. I have only had it once, in Japan of all places, and I didn't mind it. The main reason that it stuck in my mind was a 'save the bees' campaign the brand ran in 2017 that is a classic example of how bee conservationists shoot themselves in the foot over and over again.

The campaign was pretty good. They removed their mascot from the front of the box and offered packets of bee-friendly plant seeds, asking people to plant some extra forage in their gardens to bring back the bees. The company also did larger scale bee-friendly plantings with the farmers that supplied them. So far so good, right?

The Cheerios campaign was shared on social media and went a little bit viral. Problem was, the backlash from a small but influential group of bee conservationists was even more viral — particularly the accusation that some of the plant species in the packets were invasive weeds in some areas. It was a mistake, for sure, but a fairly small one; the offending plants were on sale basically everywhere in the United States. This didn't matter to the people piling on, and the backlash kept growing on their social

Edible Nasturtium flowers in a planter box.

Sunflowers — spectacular and non-invasive.

media pages about Cheerios spreading 'weeds.'

The basic problem was that in a big country like the United States, almost any seed mix will contain a species that is a problem somewhere. To truly address this would require a specific mix for each region, and native seeds to satisfy the 'natives or nothing' crowd. Not impossible, but quite difficult and hard to organize.

Eventually the initiative ended and was not repeated. The company gave out seeds, but got beaten up on social media for its trouble. I am sure that the nay-sayers thought that they had a win: 'That'll teach 'em, these folks glorifying the exotic honeybee over my beloved native bees.' But it wasn't the win they thought it was.

You see, at this time I was working inside a large, multinational corporation as they were starting to get interested in sustainability and innovation. They had had some positive success with a range of fun, innovative projects, but it was nothing compared with the way the office beehives had captured attention. People loved them and they were keen for more. However, the controversy over the Cheerios seeds had a chilling effect on this. Internal PR instantly became more cautious (and rightly so). They realized that this was becoming a polarizing issue as bees swelled in popularity, and that basically no matter what they did, someone was going to jump up and down about how they were doing it wrong.

A healthy dose of caution is not a bad thing, and both 'greenwashing' and 'beewashing' are real issues in the media, but it still bothers me how quickly this turned negative. A fun, enthusiastic but imperfect initiative wasn't helped along with constructive criticism. It was crushed, and the momentum for other companies to try their hand at conservation was crushed with it, at least for a little while.

I feel this way every time I see bee enthusiasts jump on a positive story and start picking it apart. They think they are fighting for righteous truth, but in reality they are kicking an own goal, over and over again. I wish I could reach out to these people, perform a Vulcan mind meld and get them to see the issue from a non-bee nerd perspective. I would say, 'Let them have this, let them take joy in a little bit of nature returning to the city, let it light a spark, a hunger for ever more nature around their homes.'

The brutal reality is that most people don't care about insects at all. Getting them to not see bees as a stinging nuisance is a much bigger win than most ecologists realize, given that most hardly ever do outreach to the public. Social media is geared towards negative pile-ons, and almost every initiative you can think of can be improved somehow. If something is clearly 'greenwashing' then by all means give it a kick, but let genuinely good ideas flourish. Maybe even get involved to make it better.

Bulbine Lilies — a beautiful, delicious native vegetable, but a hard-to-source seed.

A honeybee enjoying some Melaleuca flowers.

LIVING WITH EACH OTHER (AND THE BEES)

After a bunch of really tough years, it was the season of 2021/22 when things started to come good for my hives. The honeybee colonies at Parliament and on city rooftops that I had nursed and cared for burst into life, producing a mass of honey, and the native bee hotels were full of colourful, ingenious native bees, wasps and beetles.

The community of beekeepers I had been helping get established within local embassies also burst to life, with ambassadors from over seven countries donning bee suits, with more on the way. Charity events and a revitalized World Bee Day let us finally dream of a post-Covid world. The problems hadn't gone away, of course — there were still occasional kills of colonies from pesticides in the suburbs, and the threat of climate change continued to loom in the background, but the community was also banding together.

Bees have a nice way of keeping us humble, too. Just when you think you have it all figured out, they will do something completely unexpected and make you feel like a total dunce. There really are still several lifetimes of discovery to be had in our buzzy, essential friends, and if we want those extra lifetimes for ourselves and our kids, we need to learn to get along with them. For all of our fancy technology, our food supply is still largely reliant on a bunch of bugs. I genuinely hope that this never changes, and that we just learn to care for them, and about them. And care for each other at the same time.

Parliament stingless bees filling their honey pots.

World Bee Day film night.

Parliament apiary.

Endnotes

Chapter 1: Bees in the city

1. US General Services Administration, 'Pollinator Protection Initiative', www.gsa.gov/real-estate/design-and-construction/urban-development-good-neighbor-program/pollinator-protection-initiative.

Chapter 2: Meet the bees

1. Wheen Bee Foundation, 'Ensuring our food security', www.wheenbeefoundation.org.au/about-bees-pollination/.

2. Prendergast, K.S., Tomlinson, S., Dixon, K.W., Bateman, P.W. and Menz, M.H.M., 2022, 'Urban native vegetation remnants support more diverse native bee communities than residential gardens in Australia's southwest biodiversity hotspot', *Biological Conservation*, www.sciencedirect.com/science/article/abs/pii/S0006320721004602.

3. Atlas of Living Australia, 'Metallic Green Carpenter bee', https://bie.ala.org.au/species/urn:lsid:biodiversity.org.au:afd.taxon:09992614-4f27-4e08-bb48-53bb1daf6e19#overview.

4. Taxonomy Australia, 'This big, beautiful bee is in serious trouble', https://www.taxonomyaustralia.org.au/post/this-big-beautiful-bee-is-in-serious-trouble.

5. Bee Aware, 'Industry', https://beeaware.org.au/industry/.

Chapter 4: A place to bee at home

1. Goulson, D. 2019, 'The insect apocalypse and why it matters', *Current Biology*, https://www.cell.com/current-biology/pdf/S0960-9822(19)30796-1.pdf.

Chapter 5: Planting for bees

1. Plant, L., Rambaldi, A. and Sipe, N., 2016, 'Property value returns on investment in street trees: A business case for collaborative investment in Brisbane, Australia', Discussion Paper no 563. School of Economics, University of Queensland, https://www.uq.edu.au/economics/abstract/563.pdf.

2. Science Daily, 'The forest as a shelter for insects in warmer climates?' https://www.sciencedaily.com/releases/2022/05/220506151438.htm.

3. Secret Sydney, '6 stunning spots to see Sydney's blooming Jacaranda trees', https://secretsydney.com/jacaranda-trees/.

4. Leech, M., 2012, *Bee Friendly: A planting guide for European honeybees and Australian native pollinators*, Rural Industries Research and Development Corporation, Canberra, p.184.

5. Leech, M., 2012, *Bee Friendly*, p. 180.

6. Leech, M., 2012, *Bee Friendly*, p. 232.

7. Australian National Herbarium, 'Growing native plants', https://www.anbg.gov.au/gnp/interns-2010/xanthorrhoea-glauca.html.

8. New, T.R., Sands, D.P.A. and Taylor, G., 2020, 'Roles of roadside vegetation in insect conservation in Australia' *Austral Entomology*, https://onlinelibrary.wiley.com/doi/full/10.1111/aen.12511.

9. International Union for Conservation of Nature (IUCN), www.iucn.org/sites/dev/files/local_authorities_guidance_document_en_compressed.pdf.

Chapter 9: What the hell is that?

1. University of Melbourne, 'Feral honeybees key to controlling deadly parasite', https://pursuit.unimelb.edu.au/articles/feral-honeybees-key-to-controlling-deadly-parasite.

2. Peck, T. and Seeley, T.D., 2019, 'Mite bombs or robber lures? The roles of drifting and robbing in *Varroa destructor* transmission from collapsing honey bee colonies to their neighbors', *PLos One*, https://journals.plos.org/plosone/article?id=10.1371/journal.pone.0218392.

3. Guichard, M., Dietemann, V., Neuditschko, M. and Dainat, B., 2020, 'Advances and perspectives in selecting resistance traits against the parasitic mite *Varroa destructor* in honey bees', *Genetics Selection Evolution*, https://gsejournal.biomedcentral.com/articles/10.1186/s12711-020-00591-1.

4. Russo, R.M., Liendo, M.C., Landi, L. et al., 2020, 'Grooming behavior in naturally Varroa-resistant *Apis mellifera* colonies from north-central Argentina', *Frontiers*, https://www.frontiersin.org/articles/10.3389/fevo.2020.590281/full.

5. Bee Aware, 'Small hive beetle', https://beeaware.org.au/archive-pest/small-hive-beetle/#ad-image-0.

6. NSW Department of Primary Industries, 'Braula fly', https://www.dpi.nsw.gov.au/__data/assets/pdf_file/0006/176658/Braula-fly.pdf.

7. Prendergast, K.S. and Ollerton, J., 2022, 'Impacts of the introduced European honeybee on Australian bee-flower network properties in urban bushland remnants and residential gardens', *Austral Ecology*, https://onlinelibrary.wiley.com/doi/epdf/10.1111/aec.13040.

8. Prendergast, K.S., Dixon, K.W. and Bateman, P.W., 2021, 'Interactions between the introduced European honey bee and native bees in urban areas varies by year, habitat type and native bee guild', *Biological Journal of the Linnean Society*, https://academic.oup.com/biolinnean/article/133/3/725/6211042.

9. Stevenson, P.C., Bidartondo, M.I., Blackhall-Miles, R. et al., 2020, 'The state of the world's urban ecosystems: What can we learn from trees, fungi, and bees?' *New Phytologist Foundation*, https://nph.onlinelibrary.wiley.com/doi/10.1002/ppp3.10143.

10. Prendergast, K.S., 2023, 'Native flora receive more visits than exotics from bees, especially native bees, in an urbanised biodiversity hotspot', *Pacific Conservation Biology* (CSIRO Publishing), https://www.publish.csiro.au/PC/PC22033.

11. Prendergast, K.S., 2023, 'Checking in at bee hotels: Trap-nesting occupancy and fitness of cavity-nesting bees in an urbanised biodiversity hotspot', *Urban Ecosystems*, https://link.springer.com/article/10.1007/s11252-023-01381-5.

12. Saunders, M.E., Goodwin, E.K., Santos, K. et al., 2021, 'Cavity occupancy by wild honey bees: Need for evidence of ecological impacts', *Frontiers in Ecology and the Environment*, https://esajournals.onlinelibrary.wiley.com/doi/10.1002/fee.2347.

Chapter 10: Swarm catching for fun and profit

1. Seeley, T.D., 2010, *Honeybee Democracy*, Princeton University Press, pp. 67–72.

Chapter 11: It's not about the honey

1. Wood, T.J., Michez, D., Paxton, R.J. et al., 2020, 'Managed honeybees as a radar for wild bee decline?' *Apidologie*, https://doi.org/10.1007/s13592-020-00788-9.

2. Cruz, S.M. and Grozinger, C.M., 2023, 'Mapping student understanding of bees: Implications for pollinator conservation', *Conservation Science and Practice*, Volume 5, Issue 3, https://conbio.onlinelibrary.wiley.com/doi/full/10.1111/csp2.12902.

References

Appenfeller, L.R., Lloyd, S., and Szendrei, Z. 2020, 'Citizen science improves our understanding of the impact of soil management on wild pollinator abundance in agroecosystems', *PLoS ONE*, 15(3): e0230007, https://doi.org/10.1371/journal.pone.0230007

Chicago Tribune editorial board 2017, 'A bumbled campaign: General Mills gets stung for its attempt to save bees, *Chicago Tribune*, www.chicagotribune.com/opinion/editorials/ct-bees-wildflower-seeds-cheerios-0327-20170323-story.html

Fears, D. 2017, 'The great bee bumble: Cheerios wanted to help. Its plan went terribly wrong', *Washington Post*, www.washingtonpost.com/news/animalia/wp/2017/03/30/the-great-bee-bumble-cheerios-wanted-to-help-its-plan-went-terribly-wrong/

Galluci, M. 2017, 'Not everyone's thrilled Cheerios gave away 1.5 billion wildflower seeds to save the bees', Mashable, https://mashable.com/2017/03/19/cheerios-bee-campaign-controversy/

Graystock, P., Ng, W.H., Parks, K. et al. 2020, 'Dominant bee species and floral abundance drive parasite temporal dynamics in plant-pollinator communities', *Nat Ecol Evol*, https://doi.org/10.1038/s41559-020-1247-x

Wenzel, A., Grass, I., Belavadi, V.V. and Tscharntke, T. 2020, 'How urbanisation is driving pollinator diversity and pollination: A systematic review', *Biological Conservation*, p. 241.

Wilkinson, S., Zalejska-Jonsson, A. and Ghos, S. 2019, 'Trees can add $50,000 value to a Sydney house, so you might want to put down that chainsaw', https://theconversation.com/trees-can-add-50-000-value-to-a-sydney-house-so-you-might-want-to-put-down-that-chainsaw-122710

Winstanley, G., 1649, 'The True Levelers Standard: Beginning to plant and manure the waste land upon George-Hill, in the parish of Walton, in the county of Surrey', https://scholarsbank.uoregon.edu/xmlui/handle/1794/863

Wood, T.J., Michez, D., Paxton, R.J. et al. 2020, 'Managed honeybees as a radar for wild bee decline?' *Apidologie*, https://doi.org/10.1007/s13592-020-00788-9

Index